Peacemaking
among
Primates

Peacemaking among Primates

Frans de Waal

Harvard University Press
Cambridge, Massachusetts
London, England
1989

This book is printed on acid-free paper, and its binding materials
have been chosen for strength and durability.

Library of Congress Cataloging-in-Publication Data

Waal, F. B. M. de (Frans B. M.), 1948-
   Peacemaking among primates.

   Bibliography: p.
   Includes index.
   1. Primates—Behavior.   2. Reconciliation in animals.
I. Title.
QL737.P9W28   1989   599.8'0451   88-11067
ISBN 0-674-65920-1 (alk. paper)

*For my parents, my five brothers,
and Catherine—with all of whom
I have had countless reconciliations*

Contents

5 Bonobos   171

6 Humans   229

Bibliography   273

Index   287

Acknowledgments

In 1975, under the auspices of the University of Utrecht, I began a postdoctoral study of the unique chimpanzee colony at Arnhem Zoo in the Netherlands. I am most grateful to Jan van Hooff, professor of animal behavior, who gave me abundant advice and encouragement and with whom I discussed every new observation. I supervised an average of four graduate students per year, a total of twenty-three individuals. Special thanks are due the students who helped to document the dramatic events that occurred in 1980—Fred van Eeuwijk, Tine Griede, Marion van de Klashorst, and Gerard Willemsen—and the animal caretakers—Jacky Hommes, Loes Offermans, and Monika ten Tuynte. I am greatly indebted to the Arnhem Zoo and to its director, Anton van Hooff, for allowing me to work with the chimpanzee colony there. The fact that Jan and Anton van Hooff are brothers obviously facilitated cooperation between zoo and university. My study was financially supported by the University of Utrecht's Research Pool and by the Dutch Organization for the Advancement of Pure Research.

One day in the fall of 1981 Robert Goy, director of the Wisconsin Regional Primate Research Center of the University of Wisconsin, welcomed me at the Madison airport for what was intended as a one-year stay. I am extremely grateful for his support and appreciation of my work, and for the warm hospitality extended by him and his wife, Barbara. Seven years later I am still working at the center, which offered me a staff position to study the behavior of group-living monkeys. My assistant, Lesleigh Luttrell, has become indispensable by virtue of her efficiency, reliability, and commitment to our scientific objectives. She observes the monkeys on a daily basis, maintains the computer records, and shares with me the joy of following the eventful lives of over one hundred individuals, whom we discuss as if they were family. Our research team has included, at one time or another, Kim Bauers, Maureen Libet, Katherine Offutt, RenMei Ren, and Deborah Yoshihara, and I am deeply appreciative of their contributions and enthusiasm.

The convenience of having a photography department at the center is inestimable. Bob Dodsworth developed my films and did the darkroom work for the pictures in this book with his usual high standard of professionalism. Mary Schatz and Jackie Kinney cheerfully typed the manuscript and its never-ending revisions; I thank them for these and numerous other secretarial services. Finally I am indebted to the library personnel, the animal-care and veterinary staff, the computer programmers, and other center employees on whose excellent services the scientists depend. My studies in Madison are financially supported by the National Science Foundation, and by a grant of the National Institutes of Health to the Wisconsin Primate Center.

In 1983 I traveled to California to observe the world's largest collection of captive bonobos. I am grateful to the San Diego Zoological Society for allowing me to carry out this study, and to the National Geographic Society for supporting it. I thank my colleagues in San Diego for their cooperation, especially Diane Brockman and Kurt Benirschke. The animal-care staff offered all the help I could wish, and their friendship made my stay particularly pleasant: Gale Foland, Mike Hammond, Fernando Covarrubias, and Joe Kalla. Back in Madison, Katherine Offutt assisted with the data processing.

Unique photo opportunities have been provided by Stephen Suomi and Peggy O'Neill, who kept an outdoor group of rhesus monkeys in the Wisconsin countryside, and by Ronald Noë, who introduced me to the olive baboons of the Uaso Ngiro Baboon Project, near Gilgil, Kenya. Recently I worked with chimpanzees again in a study of food-sharing behavior at the Field Station of the Yerkes Regional Primate Research Center of Emory University, Atlanta, Georgia. Several photographs and anecdotes from this period have been included in the book, although data analysis is still in progress. The research was made possible by the Harry Frank Guggenheim Foundation, and by a grant of the National Institutes of Health to the Yerkes Primate Center.

I took all the photos in this volume with half-automatic Minolta and Nikon equipment—mostly on Kodak Tri-X pan film exposed at 800 ASA, using lenses from 50 to 400 mm. The single exception is the picture on page 178, which is a reproduction from A. Portielje and S. Abramsz, *Het Artisboek* (Zutphen: van Belkum, 1922), page 125; courtesy of the Royal Zoological Society, *Natura Artis Magistra*, Amsterdam.

The book has benefited tremendously from the input of many people. For years my mother scanned Dutch newspapers for the word *verzoening* (reconciliation); I owe a great many human anecdotes to her. I have used personal communications from Otto Adang, Curt Busse, Ivan Chase, Verena Dasser, Jeffrey Dreyfuss, Wulf Schiefenhövel, Fred Strayer, Andres Treviño, and Christian Welker. I am grateful to Barbara Smuts for her thorough reading of the entire manuscript and her many insightful reflections. David Goldfoot, Jane Hill, and Lesleigh Luttrell each commented on the manuscript from very different perspectives. I also thank Vivian Wheeler of Harvard University Press for an excellent job of editing and polishing the text.

The last, or rather the first, critical reader was my wife, Catherine Marin. Quickly bored by scientific jargon, yet entertained by my primate stories, her comments on each day's production helped to shape the style of this book. I cannot imagine my life without our mutual love and support.

M   Chimpanzee (*Pan troglodytes*)   F

M   Bonobo (*Pan paniscus*)   F

M   Rhesus monkey (*Macaca mulatta*)   F

M   Stump-tailed monkey (*Macaca arctoides*)   F

Aggression can certainly exist
without its counterpart, love,
but conversely there is
no love without aggression.

—*Konrad Lorenz*

Prologue

Fires start, but fires also go out. Obvious as this is, scientists concerned with aggression, a sort of social fire, have totally ignored the means by which the flames of aggression are extinguished. We know a great deal about the causes of hostile behavior in both animals and humans, ranging from hormones and brain activity to cultural influences. Yet we know little of the way conflicts are avoided—or how, when they do occur, relationships are afterward repaired and normalized. As a result, people tend to believe that violence is more integral to human nature than peace.

Ethologists delivered that pessimistic message in the sixties, and they certainly have not countermanded it in the seventies and eighties. The predominant approach in biology has been to look at life as the "continuous free fight" or "gladiators' show" that Darwin's public defender, Thomas Henry Huxley, a century ago proclaimed it to be. The focus is on ruthless competition and the benefits animals extract from their dealings with others. That animals are in a struggle for existence is undeniable; they can be amazingly violent when there is a conflict of interest. But not everything they do is at the expense of others. Many species unite in cooperative groups, which most of the time give the impression of harmony.

Our closest relatives, the primates, form stable social relationships. The members of a group are simultaneously friends and rivals, squabbling over food and mates, yet dependent on one another and having a strong need for comforting body contact. These animals have to face the fact that sometimes they cannot win a fight without losing a friend. The answer to this dilemma is either to reduce competition or to repair the damage afterward. The first solution is known as *tolerance*, the second as *reconciliation.* Nonhuman primates, familiar with both, sustain their communities by a highly developed cooling system that prevents overheating, explosion, or disintegration of the social machinery. They act like human families, many of which manage to cohere for twenty years or more in spite of being veritable battlegrounds.

Because my research concerns the principles of peaceful coexistence, I focus on sharing rather than on competition, and on how fights end instead of how they start. Reconciliation is crucial: immediately after a fight the two adversaries tend to stay away from each other, but after a time one approaches the other and tries to make friendly contact. The length of the process varies; whereas monkeys generally make up within minutes, humans can take days, years, even generations to do the same. Thus I followed with fascination reports on the meeting in a prison cell between Pope John Paul II and Mehmet Ali Agca, during which the pontiff tenderly held the hand of his would-be assassin and forgivingly talked to him ("I spoke to him as a brother, whom I have pardoned, and who has my complete trust"). Most commentators saw this as a demonstration of Christian forgiveness, but I recognized deeper roots, comparing the scene to reunions in the primate groups that I have studied.

In treating peacemaking behavior, I am one biologist who endorses the "Seville Statement on Violence" with substance as well as with spirit. This watershed document, released in 1986, resulted from a meeting of international experts on aggression in Seville, Spain. Not that I totally agree with the statement. In order to reach their conclusion that "biology does not condemn humanity to war," the authors chose to downplay our evolutionary heritage. My own support comes from within biology, complementing instead of denying previous insights. I view aggressive behavior as a fundamental characteristic of all animal and human life, but I also believe that this trait cannot be understood in isolation from the powerful checks and balances that evolved to mitigate its effects.

Although I draw parallels between animal and human behavior, even at the level of international politics, I am not in search of an animal model for our species. Each organism deserves attention for its own sake, not as a model for another. The word "primates" in the title of this book covers the human race as one of approximately two hundred primate species, all of which should be treated as equal. This means that both similarities and differences are of interest, and that no comparison is off limits. So if one accepts extrapolations from, say, the behavior of rhesus monkeys to the behavior of chimpanzees, there is no reason to object to similar comparisons between humans and chimpanzees—especially if one considers that the latter pair shares more biological traits than the former.

When we make such comparisons, it is very important that we consider humans in the same way that we consider monkeys and apes;* there is no need to place ourselves on the usual pedestal. Furthermore, it is vital that our judgments not be moralistic; "good" and "bad" are adjectives all too easily applied in this arena. Such evaluations hamper objective analysis. Even though aggression is integral to every social relationship, our inclination to denounce it is so strong that we sometimes question whether aggression should be counted as social at all.

This misconception results in part from an equating of aggression with violence. But violence is only the most extreme, not the most common, expression of aggression. Another reason for denying its social import is that aggression is not always obvious. Antagonism can be buffered so effectively that, at the surface, all we see is peace and harmony. Georg Simmel, a social philosopher of the turn of the century, pointed out that societies are not built on pure friendliness. In order to achieve a definite organization they require both attraction and repulsion,

---

*The word "apes" is not synonymous with the word "monkeys." Gorillas and chimpanzees belong to the apes; baboons and macaques to the monkeys. Apes have no tail and are larger than monkeys. They have a broader chest and longer arms with great rotational capability at the shoulders. Because humans share with apes the characteristics that distinguish them from monkeys, humans and apes are classified together in the hominoid superfamily.

integration and differentiation, cooperation and competition. Conflict and its resolution serve to overcome these dualisms and reach some form of unity. Simmel saw the peaceful termination of social struggles as a special form of synthesis—a higher process that includes both union and opposition.

In 1963 Konrad Lorenz, the father of ethology, published his well-known *On Aggression*. The book's original title spoke of a so-called evil, *Das sogenannte Böse*. The German word "sogenannt" is, more than its English equivalent, used to question the correctness of a label. Hence, Lorenz's title implied that aggression might not be as evil as commonly thought. While his book did treat the relevance of aggression in connection with love and affection, these important reflections were lost in its main message—that humans do possess a killer instinct, and they unfortunately lack the inhibition to control it.

This thesis generated a huge controversy, especially in the Anglo-Saxon world. The result was contradictory books which usually dwelt on the few gentle, nonmartial human societies that have managed to survive in remote corners of the world. In addition to this search for exceptions, the critics called on comparisons with our closest relatives. Because in those days the great apes were known as peaceable vegetarians, their Rousseauistic life-style was used as an argument against Lorenz's depiction of human nature. Ironically, these extrapolations were made by precisely the category of scientists who normally resist any comparison between humans and animals. In the face of newer data, these opponents must now feel more than a little foolish. In the past decade we have learned from field studies by Dian Fossey, Jane Goodall, Toshisada Nishida, Akira Suzuki, and others that gorillas and chimpanzees kill members of their own species. We also know that wild chimpanzees occasionally hunt, eat meat, and are cannibalistic.

The aggressive nature of humans is undeniable. We need only turn on the television set at news time, or read the history books of any nation, to get the proof and the details. Therefore, the question is not how to eliminate aggression from this world—a hopeless enterprise—but how to keep aggression under control. People value their relationships to the extent that they maintain them in spite of rivalries and disagreements. It is time for us to seriously investigate the natural mechanisms of conflict resolution. For as a result of these mechanisms, aggressive behavior is not always disruptive, and there is both a destructive and constructive side to interpersonal conflict.

I first realized that this subject can be studied in other species after witnessing a fight in the chimpanzee colony of the Arnhem Zoo in the Netherlands. It was the winter of 1975 and the colony was kept indoors. In the course of a charging display, the dominant male attacked a female, which caused screaming chaos as other chimpanzees came to her defense. When the group finally calmed down, an unusual silence followed, with nobody moving, as if the apes were waiting for something. Suddenly the entire colony burst out hooting, while one male worked the large metal drums in the corner of the hall. In the midst of the pandemonium I saw two chimpanzees kiss and embrace.

Strange as it may sound, it took me hours to realize what had happened. I kept thinking about the embrace and the excited group response. It seemed more than a mere sequence of interesting behavior patterns: the embracing individuals had been the same male and female of the initial fight. When the word "reconciliation" popped into my mind, it immediately illuminated the connection. From that day on I noticed that emotional reunions between aggressors and victims were quite common. The phenomenon became so obvious that it was hard to imagine that it had been overlooked for so long by me and by scores of other ethologists.

The Arnhem Zoo keeps the world's largest chimpanzee colony. For fear of violence, few zoos or institutions dare to build groups of this size. Many zoos exhibit gorillas and orangutans in spacious enclosures while still housing their chimpanzees in old-fashioned cages. From the time of its establishment in 1971, all went well with the Arnhem colony until 1980, when two males banded together to eliminate a rival. That bloody incident profoundly influenced my thinking about conflict resolution. Until it took place I had looked at reconciliation from a rather romantic standpoint. Now I have a very sober picture of what may happen if conflicts are resolved the hard way.

With this in mind, and realizing that the topic of reconciliation requires study of more than one species, I established projects in the United States on bonobos and on two different species of macaques—rhesus monkeys and stump-tailed monkeys. Bonobos are supposedly gentle apes, whereas rhesus monkeys have the reputation of being about the nastiest, most intolerant primates alive. I share this opinion, but have come to like these little rascals nonetheless; I see it as a special challenge to demonstrate that rhesus monkeys too have developed adequate forms of peacemaking. There simply is no other way for any animal that prefers group life to a solitary existence.

After many years of carrying out such research, I decided to communicate my findings to a larger audience—a decision that was not hard to make but was certainly not free of risks. It is virtually impossible to satisfy fellow scientists and at the same time interest the layperson. Because I have opted here for the general reader, I will make my points mainly with descriptions, anecdotes, and the photographs I have taken during my various projects. Skepticism toward this sort of evidence is understandable. Any researcher normally wants to see statistics before accepting another investigator's claims. I apply the same criteria to my own work.

Ethologists base their conclusions on observable behavior, following rigorous methods of data collection. For a certain action to be classified as "aggressive," for instance, the action has to include several specific behavior patterns that previous analyses have shown to be associated with chasing and biting. Subjective determination of the meaning of the action is thus eliminated. I have consistently followed these procedures: for every anecdote in this book hundreds of records have been entered into our computer. Readers wishing to make their own judgments are referred to the Bibliography for a listing of my technical articles.

My main purpose here is to correct biology's bleak orientation on the human condition. In a decade in which peace has become the single most important public issue, it is essential to introduce the accumulated evidence that, for humans, making peace is as natural as making war.

CHAPTER ONE

False Dichotomies

It is scientifically incorrect to say that in the
course of human evolution there has been a
selection for aggressive behavior more than for
other kinds of behavior.
> —"Seville Statement on Violence"

Mine is a policy of peace. It is founded not upon
words, gestures, and mere paper transactions,
but comes from an elevated national prestige
and from a whole network of agreements and
treaties which cement harmony between
peoples.
> —Benito Mussolini

Three unrelated dichotomies have dominated research on aggression. First, there is the classification of certain behaviors as desirable and others as undesirable. Second, over the past decade biologists have emphasized the individual at the expense of the social group. Simmel's ideas concerning the role of conflict at the societal level have been overshadowed by the view that conflict merely serves the interests of the winning party. Finally, there is the distinction between studies on animals in their natural habitat and on those in captivity. Whereas some scientists regard field research as the only work that matters, others compare such research to an uncontrolled, inconclusive experiment. Each of these dichotomies has its usefulness, but in this chapter I will challenge them all. For I strongly believe in the complementarity of different concepts and approaches.

"Good" Aggression

Two village heads of the Eipo-Papuans were about to make their first trip in an airplane. They had helped build an airstrip in the inaccessible highlands of New Guinea, and in return were invited on the flight by Wulf Schiefenhövel, a German ethologist, who told me this story. The two Papuans, not at all afraid to board the plane, made a puzzling request; they wanted one side door to remain open. Wulf explained to them that it is cold up in the sky, and that they, being absolutely naked except for their traditional penis sheath, would freeze. The men answered that they did not care. Next, they expressed the desire to bring along a couple of heavy stones. "Why would you want to do that?" asked Wulf in astonishment. The reply was that if the pilot would be so kind as to circle over the enemy village, the men could shove the stones through the open door out of the plane. Obviously their request was not granted. At the end of the day the scientist could note in his diary that he had witnessed the invention of the bomb by neolithic man.

The mind of *Homo sapiens* evidently follows the same dark path everywhere. At the same time, most of us claim to be peaceloving. In order to understand this paradox, we need to make a distinction between in-group and out-group. All human societies distinguish between killing within their own community, an act that is judged and punished as murder, and killing outsiders, which is often seen as an act of bravery and a service to the community. The lack of inhibition described by Lorenz applies, as far as I can see, mainly to war and other forms of aggression between people of different communities. If this were not so, it would be difficult to explain the cohesiveness and complexity of human societies. A gang of uncontrolled killers would build a very different kind of society indeed. Such a society might fit George Myers' description of cold-blooded terror in a school of piranhas: "The fishes swam slowly about, each keeping well away from its fellows and showing a plain desire not to have another directly behind it, where the neighbor could attack unseen. They reminded me of a set of ruthless gunmen, each with a pistol in his pocket and each quite mindful that all the rest were ready to use them at any moment."

The evolution of safety measures against damaging aggression started with the care of offspring. Even crocodiles, archaic animals with powerful jaws, may be seen walking around with a mouthful of trusting youngsters, who look out from between their mother's teeth like sightseers from a bus. The more complex the group life of animals becomes, the more remarkable the inhibitions that can be observed, not only toward kin but also toward unrelated members. Nonhuman primates are endowed with particularly highly developed checks on escalated fighting. Some of these are innate, others seem publicly enforced. Severe attacks by young adult males on females, for example, are often stopped by other members of the group. Older males have learned to control their aggression against females.

Similar social rules and acquired inhibitions play a role in human social life. If a woman hits her husband in public, this is not nearly as socially disturbing, except to the poor man himself, as when it occurs the other way around. In the first case we may think "What a temperament!" whereas in the second case we tend to think disapprovingly "What a brute!" I recall a Peanuts cartoon in which Lucy nastily smiles in Charlie's face and challenges him: "You can't hit me, Charlie Brown! I'm a *girl!*" Because of the difference in physical strength, lack of respect by men for women is a serious matter. In the privacy of our homes the fighting rules between the sexes do not always conform to the ideal, as is now increasingly evident. Male aggression may tip over into violent crime if the appropriate balancing mechanisms and social controls are lacking. Undoubtedly, the degree to which men are in charge of their anger depends very strongly on the education and example that they receive from society as boys.

The most direct way of avoiding escalation is via soothing remarks or body contact. Tension regulation by means of gentle touching, grooming, or embracing exploits the insatiable need for contact that is characteristic of the primate order. Lorenz studied primarily fish and birds, but have you ever tried to calm an upset fish or bird? When my tame jackdaws panic they do not want to be touched at all. Preening, especially of the neck feathers, does have a tranquilizing effect, but only after disappearance of the danger. Primates, in contrast, make body contact when upset and often relax after some grooming or hugging. Young monkeys are carried by their mother for nearly a year, and chimpanzee children for up to four years. It is not surprising, therefore, that they retain the need for contact comfort all their lives. Adult chimpanzees of twenty years or more still show the clasp reaction of infants, holding firmly onto each other and screaming in times of danger or of tense confrontations with rivals. Frightened soldiers at the front are said to act in the same way.

In a series of experiments William Mason was able to demonstrate that distress caused by pain can be neutralized if people take young chimpanzees in their arms. This sounds so logical that one wonders whether we needed experimental proof. But Mason's investigation took place during a period in which human and animal behavior, at least in the United States, was explained entirely on the basis of simple reward and punishment schemes. There was no attention whatsoever to more basic needs. The most outspoken representative of the behaviorist school, B. F. Skinner, saw feelings as meaningless by-products of conditioning.

Strong mother-infant bonds, found in all mammals, were explained by the rewards derived from mother's milk. According to behaviorists, that was all there was to it. Harry Harlow, founder of the Wisconsin Regional Primate Research Center, refuted this simplistic explanation by showing that the need for contact is a crucial factor, perhaps even more fundamental than the need for milk. Motherless monkey infants were given the choice between a surrogate mother made of metal wire and provided with a milk nipple, and a "mother" without a nipple but covered with soft, warm cloth. The youngsters formed a bond with the second type of surrogate, spending the day on "her" and making only brief excursions to drink from mother number one.

Harlow's pioneering exploration of what he termed the affectional system of rhesus monkeys has been and still is highly influential, although his conclusions have met with resistance. For some scientists it was hard to accept that monkeys may have feelings. In *The Human Model* (an exposition of the usefulness of humans as a model for monkey behavior!) Harlow and Mears describe the following strained meeting: "Harlow used the term 'love', at which the psychiatrist present countered with the word 'proximity'. Harlow then shifted to the word 'affection', with the psychiatrist again countering with 'proximity'. Harlow started to simmer, but relented when he realized

An adolescent chimpanzee (*right*) seeks reassurance from her mother while watching a tense conflict in the community. (Yerkes Primate Center)

that the closest the psychiatrist had probably ever come to love was proximity."

Because primates, young and old, use contact for reassurance and reconciliation, the results of aggression are not always what we might expect. Dispersal, the spreading out of individuals over a given area, is often mentioned as the predominant effect of aggression. Some older textbooks even call this the function of aggressive behavior in animals. In primates, however, major fights are followed by a wave of grooming and other friendly contacts among the members of a group. It is conceivable that with such mechanisms in operation, mild antagonism does not disturb bonds, but actually makes them stronger. Of course, aggression by itself cannot have such an effect; it first requires attraction or interdependency between the individuals.

Male hamadryas baboons enforce the unity of their harem by biting females who attempt to stray in the nape of the neck. This is followed by a "reflected escape"; instead of running away from the male, which would be the most logical response, the female runs toward him and resumes her position nearby. There is also evidence that the attachment of an infant monkey to its mother is strengthened by maternal punishment and rejection. Further, there are theories about the reinforcing effects of status-related aggression. These theories emphasize the enormous amount of attention paid to the rulers of the group, by making them either the center of visual attention, or the focus of fur cleaning by subordinates. In the same vein, it is said of dogs that they lick the hand that beats them. Obviously, this can only be true of a species with a strong sense of hierarchy. Never expect it of your cat!

If we look for a human analogue of the use of aggression in

Among wild baboons grooming is the most common friendly contact. In addition to cleaning the fur, it has a calming effect, as evidenced by the relaxed posture of this adult female olive baboon enjoying the attentions of a juvenile. (Gilgil, Kenya) bonding, a pertinent example is initiation procedures. As a young university student I myself underwent all the jokes and humiliations, including the shaving of my hair, deemed necessary for the joining of a fraternity. In those days being initiated was not entirely risk free; injuries, even death, were known to result. In this case, too, attraction is a prerequisite. Ridicule and hostility are unlikely to have any bonding effect on newcomers who have no wish to join the club. It is only in combination with desire that harsh treatment can serve both to test the new members and to strengthen their attraction and loyalty. The fact that painful initiation rituals are known from a great variety of human cultures makes it unlikely that the psychological mechanisms involved in this peculiar bonding process are an independent invention of each society.

In very global terms, then (we lack the knowledge to be more precise), it seems that aggression is often so well integrated into otherwise positive relationships that it begins to contribute to their strength. Aggressive behavior has its dangers and needs to be contained, yet it also serves to achieve solutions and compromises when there are conflicts of interest. In the absence of possibilities for open disagreement, individuals might drift apart or become insecure about one another's intentions. Aggression and subsequent appeasement, as we are learning, have an intensifying effect on relationships so that, paradoxically, some forms of abuse may tighten the social bond. In psychiatry, ambivalent but powerful attachments resulting from sexual or child abuse are not unknown.

One outdated theory suggests that anger and murderous tendencies are like water accumulating in a reservoir behind a dam. According to this "hydraulic" or "ventilationist" model, the discharge of bad feelings is both spontaneous and inevitable. I rather prefer the metaphor of aggression as fire. A pilot flame is burning in all of us, and we make use of it as the situation demands. Not in an entirely rational and conscious way, but not blindly either, as if we have to get rid of stored energy. And when things get out of control, which they regularly do, we do not exclaim that fire in itself is malevolent. We realize that it is indispensable.

The taming of fire was one of the landmarks of human history. The taming of aggression must have occurred long before that. One of the indications that primates are better at conflict management than many other animals, including the rats with which Lorenz compared us, comes from recent research on the effects of crowding. When large numbers are kept in small living spaces, rats are known to kill and even devour one another. Similar experiments with monkeys have produced much less dramatic effects. The most detailed study to date, by Michael McGuire and coworkers, compared groups of free-ranging vervet monkeys with groups kept in enclosures of different sizes. Nothing remotely resembling the bloodbath among rats occurred even under the most crowded conditions. Instead, as their living space was reduced, the vervet monkeys paid less attention to their groupmates. They looked in every direction (the sky, the ground, the outside environment) except at each other, as if they were trying to reduce the social input. This is an effective way to bypass irritation and friction, comparable to subway passengers avoiding eye contact by staring out of the windows into the darkness.

The only study of crowding in apes indicates that they go one step further than the monkeys; they actively reduce social tensions. The large colony of chimpanzees of Arnhem Zoo spends the winters in a heated hall twenty times smaller than their enormous outside enclosure. By comparing their behavior during outdoor and indoor periods, Kees Nieuwenhuijsen and I found that the increase in aggressiveness under the crowded conditions was surprisingly small. Since we also found that they groomed one another more in the hall, and that they exchanged many more appeasing greeting gestures, we speculated that these behaviors were used to minimize hostilities.

The same link between strained relationships and an intensification of contact could be observed during power takeovers among the adult males dominating the colony. Status contests always started during outdoor periods, presumably because indoors there are fewer escape opportunities—a situation that makes the challenging of an established leader very risky indeed. The extremely tense months during which status reversals are decided can easily be recognized in our graphs of the rate of grooming behavior; males never groom as much as when their position is at stake. Moreover, the maximum activity occurs between the two principal rivals. Here too we see primates coping with antagonism rather than allowing it to destroy their relationships.

"Bad" Peace

There are hundreds of definitions of aggression in the scientific literature. In English the term has a remarkably broad meaning, including such usage as "an aggressive radio reporter" or "an aggressive piano concert." Even when restricted to physical abuse or the threat of it, the term means different things to different people. Many scientists classify aggression as antisocial behavior; I am not so sure, in view of the way it is embedded in powerful buffering mechanisms that mitigate its effects.

With the word "peace" we have the opposite problem. People invariably regard peace and reconciliation as desirable goals. I intend to use some human examples to demonstrate that the word "peace" can be just as deceptive as the word "aggression." The connotations and moral values attached to these words coax us into false dichotomies, whereas in real life we rarely meet pure forms. In the absence of adequate information about human peacemaking at the personal level, I draw my examples from the only domain in which the topic is regularly discussed: international politics.

Whereas peace, generally speaking, may be good, the crucial question is good for whom? The Pax Romana must have been a blessing for the Romans, but could the same be said for all the subjects of their empire? Everybody wants peace on their own terms. That is why peaceful relationships can become unbearable for one of the parties, and why war and revolution can be seen as means to change the conditions of peace. Even the Norwegian Nobel Committee was confused by this phenomenon. While it is evident that Lech Walesa's Solidarity movement, rather than promoting harmony, was threatening the status quo in Poland, he nevertheless received the 1983 Peace Prize. In Western eyes the movement stood for a just cause; hence the curious interpretation of revolt as a peace effort.

Conor Cruise O'Brien, former editor of the *Observer*, relates how, in the fifties, a draft resolution for the United Nations required the approval of a Tibetan counselor of the Dalai Lama. The draft included the word "victory." The counselor objected to so gross a word on the grounds that his people adhered to a religion of peace. Cruise O'Brien asked whether Buddhists get involved in conflicts, and how they would describe a state of affairs which their side won. "We do have words for that," the counselor answered, "we call that very excellent best peace."

The word "peace" is the lullaby of politicians all over the world. The "War Is Peace" rhetoric of *Nineteen Eighty-Four* is recognizable in terms such as "pacification" for wiping out entire villages in Vietnam, "peacekeeping forces" for the British army in Northern Ireland, and "Peacekeeper" for a deadly missile weapon. When President Ronald Reagan selected this nice new name for the MX missile, Eugene Carroll, a retired admiral of the U.S. Navy, compared it to calling the guillotine a headache remedy.

Another misleading choice of words was the so-called Peace Movement in the Eastern bloc. This campaign purportedly shared the ideals of the strong West European peace movement. Except that the Eastern movement did not seek the disarmament of all troops—only of its Western neighbors. The governments of communist countries seemed to encourage the movement, at the same time arresting citizens who were publicly critical of arms buildup on *both* sides.

"He had chewed the word peace with the same wholeheartedness with which one masticates chewing gum," wrote the Italian journalist Oriana Fallaci about King Hussein of Jordan in

Interview with History. The king had insisted that he was trying to reach agreement with the Palestinian fighters in his country and that he would not throw them out: "I've chosen to keep the fedayeen and I keep faith with my choice. Even if my position may seem quixotic or naive." A few months after Fallaci's interview Hussein's troops made a surprise attack on the fedayeen. Thousands of them were killed, including defenseless people in refugee camps. The troops were merciless, cutting off the arms, legs, and sometimes genitals of their tied-up victims. Others were decapitated. The massacre, known as Black September, gave the king a reputation as the butcher of the Palestinians. Yet fourteen years later, in 1984, he was publicly kissed and embraced by Palestinian Liberation Organization leader Yasser Arafat. "Reconciling the Irreconcilable," as one newspaper headline put it. This dramatic peace initiative was forced on Arafat after he had lost all his strongholds in Lebanon.

Opportunistic reconciliations are to be expected in any organization in which power is decided by coalitions and group support. Chimpanzees have essentially this type of organization, albeit in a much less institutionalized form. Their leaders too make up under pressure of circumstances. The Arnhem Zoo colony was dominated for many years by a coalition of two adult males. The youngest male, Nikkie, had become leader with the help of an older male, Yeroen, who was much more experienced in intricate power games. Nikkie was physically dominant over Yeroen, yet at the same time heavily dependent on him, because there was a third male in the colony who was not afraid of either of the two ruling males individually. If Nikkie and Yeroen were in agreement, and they almost always were, there was no problem. Together they could bully the other male.

The trouble started if they got into one of their occasional fights. Nikkie and Yeroen would scream and chase each other around the large enclosure; the longer this would last, the more impressive the third male would grow. He would perform spectacular intimidation displays, hooting with his hair on end and hurling stones and branches in every direction. This male, Luit,

After a wild and noisy conflict Nikkie, a wide grin on his face, approaches his adversary, Yeroen (*right*). Yeroen raises his arm in invitation. The subsequent embrace of the two top males seals the peace in the colony. (Arnhem Zoo)

would put the group in disarray by terrorizing the females and displaying closer and closer to the two quarreling dominant males. There would be only one way to stop him: a hasty restoration of the coalition. In the midst of their dramatic conflict Nikkie would begin to make conciliatory overtures to Yeroen. He would stretch out his hand and, with a broad nervous grin on his face, beg Yeroen to make up. As soon as Yeroen had given in and accepted an embrace, Nikkie would go over to their common rival to emphasize his position. He would perform a dominance display, approaching with inflated body and lips pressed firmly together. Luit would respond with submissive bowing and grunting. He understood that a reconciliation between the other two males meant that they had formed a united front once again.

Other members of the Arnhem group also seemed thoroughly familiar with the mechanism. I have seen Mama, the oldest female, effectively mediate conflicts between the two co-alition partners. On one occasion she went first to Nikkie to put a finger in his mouth, a common gesture of reassurance among chimpanzees. While doing so, she impatiently nodded her head to Yeroen and held out her other hand to him. Yeroen came over and gave Mama a long kiss on the mouth. When she withdrew from between them, Yeroen embraced the still screaming Nikkie. After this reunion the two males, side by side, chased off Luit, who had begun to strut around, hair on end. In effect, Mama had put an end to the chaos in the group by literally repairing the ruling coalition.

Peacemaking is a complex matter, dependent on both strate-gic considerations and a desire for congenial relationships. The latter, subjective factor is sometimes glorified as the only one that counts. People love to imagine a garden in which the pro-verbial wolves and lambs cheerfully play together or where, for that matter, Russian and American soldiers exchange bouquets of flowers. According to former president Richard Nixon, the utopian type of peace is achieved at two places only—at the typewriter and in the grave. It has no practical meaning in a world in which conflict among people is persistent and perva-sive: "If real peace is to exist, it must exist along with men's ambitions, their pride, and their hatreds." A similar concept underlies former premier Nikita Khrushchev's term "peaceful coexistence." After Stalin's death the Soviets were determined to improve their international reputation. Khrushchev observed that since neither the communist nor the capitalist states want to make a trip to Mars, they will have to exist together on one planet.

Nixon and Khrushchev may not seem the kind of fellows from whom we should take lessons in peace, but it is exactly their kind on whom our future depends. Their cynical opinion, that mutual fear rather than mutual trust forms the basis of international peace, contrasts with that of many pacifists, who propose unilateral arms reduction as a solution. With their more optimistic vision of human nature, the pacifists are operating from a fundamentally different concept of peace. Although I do not share their optimism—any major power imbalance would frighten me terribly—it is equally hard to see the ongoing arms buildup as a rational undertaking. The illusion of rationality can take on ridiculous proportions in humans, when in fact all we seem to be doing is following some rather primitive action-reaction chain of escalation.

All parties involved in the arms control debate—perhaps the most important public debate ever—wish to hang the same label on their own cause. This struggle for the right to talk of "peace" both demonstrates and warns against the incredible power of the word. A British journalist, Bernard Levin, has complained about the pacifists' claim to it. "The very word 'peace' has been stolen from its honourable place in the language and used to suggest that those who believe that peace may be more easily and safely secured by strength are not seeking peace at all; indeed, much of the time the disarmers go further and use for themselves the word 'anti-war', with the clear implication that those who reject their case are 'pro-war' " (London *Times*, July 7, 1983).

It definitely is a long way from the balance between superpowers to the individual relationships of primates, which I shall be treating in this book. What the two have in common is that to describe their interactions in terms of "peace" and "aggression" is almost meaningless in the absence of information about the precise circumstances. We know fairly well what we mean by these words, but we are also used to applying qualifiers. In our own relationships we rarely confuse a peace based on trust with one based on opportunistic considerations, mutual fear, or total domination; when studying primates, we have to bear in mind the same distinctions.

The Individual and the Group

A person who has never heard of the dance of the honey bee will fail to recognize that phenomenon when watching bees.

Karl von Frisch did not see it during many years of intensive study until, in 1919, he made his far-reaching discovery. His insight brought order to the apparent chaos of the hive, forever changing the way ethologists look at bees and at animal communication in general. The perspective of an observer depends on such previous discoveries, on training, on theoretical developments, and even on the sociocultural climate of the day.

It is always useful to consider an investigator's background and frame of reference. There are globally three perspectives from which social behavior can be studied: from the standpoint of the group as a whole, the standpoint of the individual, or the standpoint of the genetic material. However odd the last approach may sound, it receives much attention in a branch of ethology known as *sociobiology.*

From a theoretical standpoint, the genetic basis of behavior is very interesting. A behavior that has been inherited from ancestors must have benefited or at least not harmed them, else they would not have survived and reproduced. Inborn traits have proved their worth in the course of millions of years of evolution. If this Darwinian view is carried to its extreme, animals as well as humans are regarded as mere "survival machines" serving the multiplication of their genetic material. A gene's future depends entirely on the reproduction of its carriers, that is, the individuals in whose chromosomes the gene finds itself. If these owners fail to leave offspring, the gene will not reach the next generation. Successful genes, according to this theory, produce behavior patterns that help their carriers to find food, attract members of the opposite sex, and raise offspring. Genes producing helpful behavior toward kin are also promoted, because relatives have many genes in common. From the gene's viewpoint, it does not really matter through which of these individuals it is procreated. An organism is regarded, then, as a robot designed to serve its genes: "They are in you and me; they created us, body and mind; and their preservation is the ultimate rationale for our existence." This extraordinary quotation comes from Richard Dawkins' book, *The Selfish Gene,* a highly readable explanation of these controversial ideas.

The sociobiological picture of animals presently dominates

the field. Yet when I see a pair of parrots tenderly and patiently preen each other, my first thought is not that they are doing this to help the survival of their genes. This is a misleading manner of speaking, as it employs the present tense, whereas evolutionary explanations can deal only with the past. Personally, I try to look at behavior from the animal's standpoint—the feelings, expectations, and intelligence which determine whether an animal acts this way or that. What does the male parrot see in this particular female; what does she see in him? This is the psychological rather than the biological origin of behavior. From my perspective, the preening of these birds is an expression of love and affection—or, to be less interpretative, the mark and measure of an exclusive bond. Obviously, this more empathic approach to animal behavior is hard to apply to slugs, frogs, or butterflies, but since my research is devoted entirely to monkeys and apes, I believe in its value. The decision making underlying much of what these animals do strikes the human observer as very familiar. Provided that it is based on intimate knowledge and translated into testable hypotheses, anthropomorphism is a very useful first step toward understanding a psychology similar to and almost as complex as ours.

A third approach to social behavior is at the group level. Until recently, biologists happily talked about animal behavior as if the good of the group or even the species could be the objective of individuals. We now realize that natural selection would make short work of animals who place the group's interest above their own. Contributions to society must bring some advantage to the contributor, either directly or indirectly. As a general rule, this is also true in human society. The bakery on the corner feeds the entire neighborhood, yet the baker does his job out of self-interest. Bakeries serve a double function: they provide society with bread, and they provide bakers with income. Sociobiologists wish to know how a particular behavior came into existence, and obviously this depends on what the actor and his relatives gain from it. Since, in their view, the benefits to society as a whole do not really matter, some sociobiologists have come to regard societies as mere abstractions.

To my way of thinking, this is quite narrow-minded. Each

individual may pursue his or her own goals, but society is more than the sum of their private enterprises. The group level can be studied as an independent reality, in the same way that some scientists study forests and others trees. When Mama restored peace by intervening in the fight between the two ruling males, Yeroen and Nikkie, the entire community profited. Such acts of mediation serve group stability; in the long run, they may keep a group from falling apart. At the same time, it would be naive to think that Mama did not have her own personal reasons for what she did, or that the two males restored their powerful coalition for the group's sake. We can recognize different functions depending on the level considered. In a sense, the comparison is between the "socialist" (collective interest) and "capitalist" (private interest) principles of social life. Whatever ideologues from the left or right tell us, every society has to strike a balance between these two fundamental principles.

For our understanding of society, why individuals contribute to its complex structure may be unimportant, yet I cannot help wondering about individual motives. Do primates build and maintain their communities in the same way as corals form oceanic reefs—that is, blindly, without a concept of the end product? Or do they, like people, have some image of their own society and the way it is (or should be) organized? This is of particular interest in connection with the topic of reconciliation; intricate societies are unthinkable without conflict resolution. Do animals ever resolve conflicts with this larger picture in mind? When one monkey group, for example, regularly confronts another in territorial disputes, does this lead to a greater willingness *within* each group to forgive and forget for the sake of group unity and strength? I would not be surprised if it did, and I think we should keep an open mind on the question of group consciousness.

In summary, in a time when many biologists place the inborn basis of behavior at the top of their list, we should realize that this is only one level of explanation, and that as far as the higher mammals, including humans, are concerned, it may not be the most significant one. Equal attention should be paid to the three complementary perspectives: the genetic evolution of behavior, the motives and experience of the individual, and the behavior's impact on society as a whole.

Captive vs. Field Studies

Throughout this book I describe serious hostilities among primates in order to make the point that peacemaking is not some superfluous form of hedonism. Just as war and peace are inseparable issues, reconciliation behavior must be seen in light of the threat of violence. Neither aggression nor peace is a stable state. There is constant motion and interaction, similar to that between the yin and yang principles of Chinese philosophy. Just as the yang having reached its climax retreats in favor of the yin, and the yin having reached its climax retreats in favor of the yang, there exists no such thing as eternal concordance in any social system. Pure peace is like an ocean without waves and tides. Pure aggression can only bring total destruction to oneself, the other, or both. It is the pendulum swing between conflict and accommodation that we observe, rather than either of the poles.

Disturbance of this dynamic equilibrium does occur, however. A particularly dramatic example serves to show that aggression among primates is a very real problem, one that needs to be dealt with at all costs. The price of unchecked escalation is unbearable. Monkeys and apes are not the winsome, humorous creatures many people believe them to be: they can and occasionally do kill one another. The example concerns animals under highly unnatural conditions, which prevented a normal solution to their problems.

In 1925 officials of the Zoological Society of London "liberated" no less than one hundred monkeys at Monkey Hill, a rockwork enclosure of 30 × 20 meters. The animals were hamadryas baboons, also known as sacred baboons, once worshiped by the Egyptians. This species is a feminist's nightmare. Males are twice the size of females and have enormous canines.

They are passionate harem holders, treating females as posses-
sions and defending them against other males. Unfortunately,
only six of the baboons in the enclosure were females. Monkey
Hill turned into a bloodbath. Males fought over females and in
the process dragged their prizes around. The captured females
sometimes did not have a chance to relax and eat for days on
end. Thirty females were added to the colony, but this did not
stop or slow the killing. Six and a half years later the few surviv-
ing females were removed. Sixty-two males and thirty-two females, over two-thirds of the original population, had died of
stress and injuries. Only a relatively calm masculine community
remained.

Solly Zuckerman, anatomist to the society, in 1932 described
the massacre in his influential book, *The Social Life of Monkeys
and Apes*. As indicated by his ambitious title, this was a time of
sweeping generalizations. Unaware that harem holding is a rare
specialization of the hamadryas baboon, not a general pattern,

The difference in size between male and female hamadryas baboons is
accentuated by the male's magnificent coat. (Arnhem Zoo)

Zuckerman speculated freely about the origin of our own society, including the human "compromise" of a monogamous pair-bond. He noted that baboon females in heat use their sex appeal to obtain certain privileges. Comparing this to prostitution, he overemphasized the sexual component of social life: "The sexual bond is stronger than the social relationship, and an adult male, unlike a female, is not owned by any individual fellow."

A whole generation of primatologists has worked to dispel some of these generalizations. Studies of other monkey species have shown, for example, that many of them maintain a cohesive social network throughout the year in spite of having only a brief season of sexual activity. The most striking contrast was provided by work on the same species of baboon as studied by Zuckerman. The British anatomist had not been an unreliable observer; his descriptions were remarkably detailed and accurate. Nor can he be blamed for the lack of knowledge in his day. As a consequence, however, he had failed to recognize fully the exceptional nature of the chaos and violence reigning at Monkey Hill. The possibility that the events might have been unnatural was mentioned only in a footnote.

In the fifties a Swiss ethologist, Hans Kummer, carefully studied a smaller, well-established colony of hamadryas baboons in Zurich Zoo, and later observed them in their natural habitat, the desert of Ethiopia. His work is now so well known in primatology that we might almost rebaptize the hamadryas baboon and call it the Kummer monkey. I myself have been strongly influenced by Kummer's insights into what he termed tripartite relationships, the way the interactions between two individuals depend on their connections with third parties. A relevant example is the mechanism that prevents males from fighting over females—precisely the mechanism that failed to develop in the London Zoo colony.

After field observations had indicated that male hamadryas baboons recognize one another's ownership of females, Kummer and his coworkers designed an experiment to test the development of this restraint. They first demonstrated that if a female was released in a cage with two males, they would fight over her. If the female was put with only one male, while the other could watch from an adjacent pen, the outcome was quite different. The female needed only a brief time with one male for the other to respect the pair-bond upon introduction into their cage. Even big, totally dominant males were inhibited from fighting. Instead, they looked at the sky, fiddled with small objects on the ground, or attentively scanned the landscape outside their enclosure, shifting their heads like baboons who have spotted something highly interesting. Kummer was never able to detect that object, however.

Such embarrassed reactions were typical among males who knew one another; fighting sometimes did erupt between unfamiliar males. Inasmuch as familiarity is the norm within a baboon troop, the respect for ownership demonstrated by this simple experiment may suffice to maintain the peace. Its impact at the group level must be tremendous, allowing for the multileveled organization that Kummer described: males and females live in harems; harems travel together in bands; and a number of bands are united into a troop of several hundred members who spend the nights together on the same sleeping cliffs. How orderly a society in comparison with the London Zoo colony! The procedure used at Monkey Hill had released the "beast" that lives inside the hamadryas baboon. Normally quite capable of group life, their veneer of civilization was torn away when they were indiscriminately thrown together in a crowd with the wrong sex ratio.

In contrast to Zuckerman's claim that "few significant differences can be seen between the broader social mechanisms of different monkeys and apes," it is now known that one can prove the "naturalness" of almost any social pattern by selecting the appropriate species. The variety is immense. A strong mother-offspring bond is found in all primates; beyond this, virtually everything exists, from monogamy to promiscuity, from despotism to egalitarianism. Nowadays when we look at humans from a biological perspective, the aim is to establish our place among the rest of the primates, considering separately both similarities to and differences from each of our closest relatives. Simplistic lists of traits that humans share with *the* primate are no longer accepted.

To understand the full range of possibilities, ethologists have for the past few decades been studying primates under all sorts of circumstances: in their natural habitat, in large zoo groups, and in laboratories. For a long time a sharp distinction was drawn between fieldwork and laboratory studies. Now the approaches are beginning to merge. Field-workers who trap animals collect blood samples and body measures before releasing them. The blood samples go to laboratories specializing in primates and help to determine, for example, genetic relatedness within wild groups. Conversely, lab researchers are familiar with the literature on free-living primates, which helps them to

A tense confrontation between two male olive baboons, a species closely related to the hamadryas baboon. Both in captivity and in the wild, primates fight. (Gilgil, Kenya)

interpret the behavior of their subjects and to design experiments relating to foods, sounds, temperatures, and other factors of the natural environment. As a kind of bridge between these two categories one finds researchers such as myself, who specialize in the observation of natural-sized groups of captive primates.

The dramatic increase in knowledge over the past decades has led to more nuanced concepts. For example, calling the family units of hamadryas baboons "harems," as I did above for convenience and vivacity, is questionable. If the term applies to any species, it is to this baboon, but even in their society the female is not mere merchandise. She does exert some choice. Christian Bachmann, in a lab study, measured the preference of females for certain males. He demonstrated that females who strongly prefer their own mate are less often abducted. Rival males seem to perceive the female's attachment and are less interested in acquiring an unwilling partner than one who is happy to leave.

When the lessons from Monkey Hill, the field observations, and the experimental results are taken together, we gain more insight into the society of hamadryas baboons than from each part separately. It is the judicious combination of different approaches that is the future of primatology. In no sense can captive studies in isolation replace research in the natural habitat, but they do complement it in one very important respect: *detail*. Knowledge about primates is now so much greater than in the twenties that many successful, harmonious captive groups have been established all over the world. These groups permit very thorough research, for years on end, with attention to all kinds of social subtleties. This is often not possible with feral primates. Some years ago an American primatologist came back from a two-year stay in the Zaire jungle with only six hours of observation on the shy, elusive bonobo. In a single winter I watched and videotaped ten members of the same ape species for three hundred hours in the San Diego Zoo. Obviously, my observations have their limitations, but so do the field data. (In recent years, I must add, there have been a few more successful bonobo expeditions to Africa.)

By recognizing the strong and weak aspects of each method, and by putting the results together like the pieces of a puzzle, we end up with a picture of the full behavioral potential of a species, including the effects of different environments.

Chimpanzees

The little creature, which I had punished for
the first time, shrank back, uttered one or
two heart-broken wails, as she stared at me
horror-struck, while her lips were pouted more
than ever. The next moment she had flung her
arms around my neck, quite beside herself, and
was only comforted by degrees, when I stroked
her. This need, here expressed, for forgiveness,
is a phenomenon frequently to be observed in
the emotional life of chimpanzees.

—Wolfgang Köhler

Henry: Why don't you kiss and make up?
Martha Jane: It is simply not that easy! You just
don't wipe away a hurt with
something as simple as a kiss. You
might, but I don't forget that easily.

—Anita Clay Kornfeld

Standing in the public observation post that overlooks
the chimpanzee island, Rianne Scholten and Brigitte Kint act as
if they are watching the apes and taking notes on their behavior. This comes as no surprise to most visitors, because the
chimpanzee research project at Arnhem Zoo is well-known in
the Netherlands from newspapers, radio, and television. In reality, however, the two students are documenting human behavior. The average visitor spends three and a half minutes
looking at the apes. Those who come alone spend more than
twice as much time as people in groups or families. The least
patient are adult men, who make the most overtures to leave
("Come on, let's go"). And it happens repeatedly that people
who do not stay for more than a few minutes walk off exclaim-
ing, "Oh, I could watch them for hours!"

The Arnhem Project

That is what *we* do. I estimate my own observation of the Arnhem colony in the period from 1975 through 1981 at about six thousand hours. During much of this time my students and I collected data, usually by speaking into cassette tape recorders. This method allowed us to keep our eyes on the chimpanzees while giving a verbal account of their doings. The problem with chimpanzees is that they do remarkably little most of the time. They move slowly, eat grass, sleep for a long while, groom one another. During all this the observer has to be there, and wait. On the other hand, when the chimpanzees do wake up and cause some social ripples, there is no way an observer can record with pencil and paper all that is going on. Trying to follow our fast-moving subjects, we rush along the water-filled moat around their island. To go on the island itself would be far too risky (chimpanzees are stronger than we are, and not always friendly), but even to walk on the other side is not without danger. I still keep a taped account of a major outburst of ape activity in which the observer's excited voice abruptly ends in a splash.

The island comprises two and a half acres. It is covered with grass, sand, and fifty tall trees, most of which are protected with electric wire against bark eating by the twenty and more chimpanzees. The group includes four adult males, ten adult females, and a growing number of adolescents, juveniles, and infants born in Arnhem. The adults come from different European zoos. Most of them are feral born and between fifteen and thirty years of age, which is not particularly old for chimpanzees. Every evening the apes enter the main building, where they are put into their night cages in smaller groups before they receive their meal. The building also contains two large halls used as winter quarters, and an observation post especially for the ethologists.

Zoo visitors are kept at a distance, so that they cannot provoke the apes by shouting, feeding, or imitation. Contrary to general belief, humans imitate apes more than the reverse. The sight of monkeys or apes induces an irresistible urge in people to jump up and down, exaggeratedly scratch themselves, and holler in a way that must make the primates wonder how this otherwise so intelligent species has come to depend on such inferior means of communication. The Arnhem exhibit is designed to present apes for study rather than for interaction. People have to learn to take the time to watch how chimpanzees behave among themselves. The colony is ideal for this purpose, because its size and composition are similar to those of small chimpanzee communities in the wild. There is obviously much more to be seen than among classically housed apes.

Research on our closest relatives is still in its infancy. If we assume that the chimpanzee's psychology and social life are about half as complex as those of humans—and this is, I am sure, a gross underestimation—we would need about half as much research on this species as we spend on ourselves to reach a comparable level of understanding. Whole armies of anthropologists, sociologists, psychiatrists, and psychologists investigate human behavior and do not yet have the final answers. How, then, can a few dozen chimpanzee experts have done more than scratch the surface?

Reconciliation and Consolation

Early scientists tried to understand the social life of animals without knowing the history of the individuals, their long-term relationships, or the kinship network of the group. Primatologists were the first to abandon this approach. They took the important step of identifying primates individually and following their lives over long periods of time. This involved giving names to the study objects. Other scientists frowned upon this development, regarding it as a threat to objectivity (it does sound different when you are collecting data on "Charlie" rather than on "a male chimpanzee"). If names have the effect of bringing animals closer to us, making them in a sense more human, this has not harmed science; tremendous new insights have resulted from individual recognition. We now realize how much it matters to primates with which of their groupmates they are dealing. Like people, animals have friends and enemies, and they certainly do not treat them in the same manner.

The individualized approach is crucial when we try to analyze peacemaking. Students of animal behavior used to speak of "arousal reduction," "appeasement," and "reassurance" if they saw animals engage in body contact during or after a disturbing event. The terminology stressed the effects on the internal state and psychological well-being of individuals. Though not incorrect, this view was incomplete. After a dispute, primates do not settle down in a random manner. The concept of reconciliation places these same reassurance gestures in the context of ongoing relationships between individuals. The need for contact after a fight specifically involves the former adversary, as this is the only partner with whom repair of the damage is possible. Animals seek not only psychological stability but also *social* stability.

Both reconciliation and its counterpart, revenge, require that the participants remember with whom they have had a fight. Just as the primates themselves need to keep mental records of their interactions with others, so too must every observer who wishes to decide whether or not certain contacts are related to past aggression. Recognizing individuals is the easiest part of the task. Chimpanzees differ so much in face, voice, gait, and psychological characteristics that it takes only a few days to recognize all the individuals in the Arnhem colony. To store a large number of events in the mind is more difficult, but recording devices can help. The goal is to look beyond the immediate context of behavior at entire action-reaction chains over periods of minutes, hours, even days.

Research has taught us that chimpanzees have memories like the proverbial elephants and are capable of planning ahead; observation of their social life suggests that they use these capacities all the time. An adult male may spend minutes searching for the heaviest stone on his side of the island, far away from the rest of the group, weighing the stone in his hand each time he finds a potentially bigger one. He then carries the stone he has selected to the island's other side, where he begins—with all his hair on end—an intimidation display in front of his rival. Since stones serve as weapons (chimpanzees throw fairly accurately), we may assume that the male knew all along that he was going to challenge the other. This is the impression chimpanzees give in almost everything they do: they are thinking beings just as we are.

Reconciliation relates to both past and future; it serves to "undo" previous events with an eye to future relationships. Depending on how much of past and future are taken into account, we may speak of a rational element in peacemaking. Because chimpanzees resemble us more than any other animal with regard to mental processes, the study of their behavior is especially relevant. They often delay their response to a particular situation, patiently waiting for the best opportunity. They also test the ground before making a social move, in the same way that they may throw small pebbles at a dead animal before touching it. As a result, events in chimpanzee society span relatively long time intervals. It requires some training to get the overview, but once this has been achieved the connections become crystal clear.

Let us follow an individual who has been involved in an aggressive encounter—which usually comprises lots of barking and little biting. Chimpanzees are among the noisiest animals in the world, making an indescribable din when chasing one another. Their fights rarely escalate to the point of damaging aggression, however. On this occasion Nikkie, the leader of the group, has slapped Hennie during a passing charge. Hennie, a young adult female of nine years, sits apart for a while feeling with her hand the spot on her back where Nikkie hit her. Then she seems to forget the incident; she lies down in the grass, staring in the distance. More than fifteen minutes later Hennie slowly gets up and walks straight to a group that includes Nikkie and the oldest female, Mama. Hennie approaches Nikkie, greeting him with a series of soft pant grunts. Then she stretches out her arm to offer Nikkie the back of her hand for a

A reconciliation sequence.

This page: Hennie (*on the right*), carrying an infant, approaches Nikkie after he has slapped her. Hennie first offers her hand to the aggressor for a hand kiss, after which the two engage in a mouth-to-mouth kiss.

Next page: Hennie then goes to Mama (*at left*), who has been watching, and nervously grins at her. Mama consoles the younger female with an embrace. (Arnhem Zoo)

kiss. Nikkie's hand kiss consists of taking Hennie's whole hand rather unceremoniously into his mouth. This contact is followed by a mouth-to-mouth kiss. Then Hennie walks over to Mama with a nervous grin. Mama places a hand on Hennie's back and gently pats her until the grin disappears.

Mama and Hennie have a very special relationship. Hennie was only two years old when she came to Arnhem, where she was more or less adopted by Mama. This meant that she received protection and reassurance from this influential female in case of trouble. In a sense, Mama has special relationships with everyone. She acts as the mother of the group, hence her name. Even the adult males, physically fully dominant, sometimes act like ape children in her presence. During one of Mama's interventions in a protracted conflict between Yeroen and Nikkie she ended up sitting with one full-grown male in each arm. They did not stop screaming, but seemed at least to have ceased fighting. Then, suddenly, Yeroen reached over to grab Nikkie by the arm. Mama found this unacceptable and chased Yeroen off. Later the two males made up by mounting, kissing, and fondling each other's genitals, after which they discharged their tensions by together briefly chasing Dandy, a lower-ranking male.

Over the years several of my students, especially Angeline van Roosmalen, Tine Griede, and Gerard Willemsen, have collected data on reconciliation behavior. According to their studies, around 40 percent of the time opponents contact each other within half an hour of their aggressive encounter. This is a high percentage in view of the size of the outdoor enclosure, which makes mutual avoidance quite easy. That these reunions are far from accidental is evident from the way the contacts differ from the usual ones. One distinct gesture is the outstretched arm and open hand, which chimpanzees use to beg for body contact. They also show more eye contact, yelping, and soft screaming when approaching former adversaries; most important, there is much more kissing in this situation.

These behavior patterns are not only special compared to average contact forms in the colony, they are also different from reassurance given by bystanders. Contact from individuals who were not involved in a specific fight is called *consolation*, as distinguished from *reconciliation*. In the above sequence Hennie first had a reconciliation with her adversary, Nikkie, then received consolation from Mama. Consolations involve more embraces than kisses, whereas the opposite holds true for reconciliations. In other words, when you see two chimpanzees engaged in a protracted kiss, chances are that they confronted each other not long before. When they only hug each other, it is more likely that the tension was caused by a third party.

Unfortunately, science is virtually ignorant of reconciliation behavior in private human relationships. One reason for the dearth of data may be that social psychologists study people in experimental settings. Since their subjects hardly know one another, they cannot but exhibit rather superficial relationships. Family therapists, in contrast, are extensively familiar with the phenomenon of reconciliation—it is their business—but because they supervise and influence the process, their experience does not concern the "natural" life of humans. Most people will agree, however, that kissing as a form of peacemaking is a characteristic that we share with the chimpanzee. This is even reflected in symbolic ceremonies such as the one in 1982 when the Argentine and British prelates of the Catholic Church exchanged a kiss of peace during a papal mass that coincided with the British invasion of the Falkland Islands.

Humans make up in a hundred different ways: breaking tensions with a joke, gently touching the other's arm or hand, apologizing, sending flowers, making love, preparing the other's preferred meal, and so on. Nonetheless, the kiss is the conciliatory gesture par excellence. Another point in common with the chimpanzee is the critical role of eye contact. Among apes it is a prerequisite for reconciliation. It is as if chimpanzees do not trust the other's intentions without a look into the eyes. In the same way, we do not consider a conflict settled with people who turn their eyes to the ceiling or the floor each time we look in their direction.

Our chimpanzee study found indications that conflicts are

less often revived after friendly body contact between the antagonists. It should be realized, though, that all our findings are
of a statistical nature. The connection between aggression and
subsequent kissing is undeniable, but exceptions do occur. It is
impossible to be absolutely sure, for any single friendly meeting
between two chimpanzees, whether they are making up or
whether they would have had this particular contact anyway,
regardless of previous disagreements. I find this uncertainty
frustrating, especially when witnessing what I think may be a
reconciliation that took hours or almost a day. At such moments
I really wish I could work with questionnaires. Question 1: "Did
you recall your morning fight with X when you kissed her this
afternoon?" Question 2: "Did you feel better afterward?"

Normally, the initiative for peacemaking is divided equally
between dominant and subordinate chimpanzees. Exceptions
occur after heavy physical aggression, such as biting. These rare
serious fights are less often reconciled by the dominants. Their
unwillingness to make up takes on dramatic proportions in the
final stages of a power struggle. I have observed five such struggles among adult males in the Arnhem colony; three resulted in
a reversal of the rank order, two in a reestablishment of previous positions. The process takes several months, involving
many intimidation displays and aggressive encounters, together with a few physical attacks. Confrontations between the
rivals alternate with emotional reunions and unusually long
grooming sessions. Yet these friendly exchanges decrease in
frequency when the end of the struggle approaches.

The male who eventually will emerge as the dominant starts
to refuse reconciliations during the last two or three weeks of
the tense period. Each time his rival approaches him or begs,
with hand held out, for contact, the future dominant turns his
back and walks away. Rejections are daily occurrences as long
as the loser does not formally submit. Status among chimpanzees is clearly communicated by the subordinate's pant grunts
and deep bowing movements. Once the loser becomes defer-
ential by regularly greeting the other with the characteristic
grunts, contact between the two is resumed and the relationship relaxes.

This mechanism of "no submission, no peace" is a form of *conditional reassurance;* that is, the dominant's reassurance of the subordinate, by means of friendly gestures, is contingent on the subordinate's reassurance of the dominant, by an acknowledgment of the inequality in status. Every hierarchy-oriented species has evolved special signals for this purpose. These signals are comparable to the military salute of soldiers to their commanders. The soldier who forgets to perform this ritual will soon find out that the mechanism of conditional reassurance is the backbone of every rank system. Its existence warns against the popular view of a rank order as a mere ladder of superiority. The situation is more complicated; rank orders bind individuals together in a pact of loyalty. As Rudolph Schenkel remarked about the wolf's status signals, "Submission is the effort of the inferior to attain friendly or harmonic social integration."

A four-year-old male has reached the weaning age. His mother interrupts his suckling by placing her hand under his chin and gently pushing away his head. The youngster's pouted lips express disappointment. (Yerkes Primate Center)

Conditional reassurance is not limited to status relationships. The first time mammals experience it fully is when their mother weans them. Chimpanzee mothers start doing so when their offspring are four years old. The female prevents her child from suckling by threatening or pushing it away, or by covering her breasts with an arm. This causes the youngster to do a lot of pouting and whimpering in a very humanlike voice, and an occasional temper tantrum results, in which the child screams and squirms as if its death were imminent. The mother offers calming body contact on condition that the youngster keep his head turned away from her nipples; otherwise she may push again. Since the mother is such an immensely important figure, on whom the weaned offspring will continue to depend for years, he cannot walk away from the problem. The received threats and rejections have to be accepted, and a new relationship develops with maternal warmth now being conditional on the offspring's behavior.*

The connection between infantile and adult behavior is fascinating to watch. Mature male chimpanzees may roll around screaming and beating the ground when a dominant rival rebuffs their reconciliatory overtures after a fight. They act like rejected children. Curiously enough, hooting (the vocalization with which male chimpanzees challenge and provoke their rivals) involves the same pouting expression as worn by a hungry infant, and the soft *huu-huu* sounds with which a hooting display begins resemble infant whining (although the adult male's voice is obviously much deeper). In short, there seems to be psychological continuity between the process of weaning and status contests among adults. This is perhaps because weaning is also about power; it is a reversal in the direction of social control between mother and child. During weaning an individ-

---

*Sometimes I wonder whether the change in Western culture from breast-feeding to complete or partial bottle-feeding has affected patterns of reconciliation. Compared to weaning a child from the breast, modern weaning is a much less physical, less traumatic affair. The lower intensity of both the mother-child conflict and the comfort needed afterward could very well affect a child's capacity to cope with rejections and relationship crises later in life.

Fully adult chimpanzees may lapse into childlike behavior when upset. This female begged for food from another female, but was rejected. She is screaming in frustration, hitting herself with spasmodic arm movements. (Yerkes Primate Center)

ual gets his first experience of dramatic change in a relationship that he needs to maintain.

Herbert Terrace, who investigated the sign-language capacities of a young chimpanzee named Nim Chimpsky, gave an account of how conditional reassurance works in human-ape relationships. Nim had learned the meaning of a large number of hand gestures that represent words in American Sign Language (ASL) and communicated in this medium with his teachers. The student was not always easy to handle, however, and sometimes needed disciplining. An effective technique was for the teacher to walk away, preferably while signing "You bad" or "I not love you." Nim responded to the threat of separation by a so-called sorry-hug routine. In the course of time this procedure lost its effectiveness, however, unless Nim's teachers postponed the hug. Sometimes this led to a tantrum by the pupil. Terrace noticed that Nim's behavior changed dramatically for the better after such outbursts, but that exceptions occurred when he was reassured too quickly.

It seems, then, that chimpanzees are extremely sensitive to the potential disturbance of their relationships; they may fear it even more than the unpleasant physical effects of aggression. This makes it possible for humans as well as conspecifics to demand changes in behavior before normalizing the relationship. If aggression is a stick, reconciliation is often a carrot.

Sex Differences

Male chimpanzees are more conciliatory than females. Based on years of systematic observation of the Arnhem colony, reconciliation occurs after 47 percent of conflicts among adult males, but after only 18 percent of those among adult females, with reconciliation between the sexes falling in between. This sex difference is as yet an unsolved puzzle. I have tried to link it to other sex differences in chimpanzees, of which there is no lack; a hacker entering our computer files on male and female behavior would never, on the basis of the data alone, guess that he is dealing with one and the same species. Of particular interest is a sex difference in cooperative relationships. Among males most cooperation seems of a transactional nature; they help one another on a tit-for-tat basis. Females, in contrast, base their cooperation on kinship and personal preferences. Both forms of mutual support permeate all aspects of chimpanzee communal life, including power relationships. A study of the power structure may, therefore, throw light on the sex difference in peacemaking.

The law of the jungle does not apply to chimpanzees. Their network of coalitions limits the rights of the strongest; *everybody* pulls strings. When two apes get into a fight, others hurry over to watch the scene, give high-pitched barks of encouragement, or intervene on behalf of their favorites. Coalitions against a single individual range in size from two to ten aggressors. But this victim may also receive help, leading to large-scale confrontations between different sections of society. Combatants actively recruit support. They draw attention by screaming at the top of their lungs; they put an arm around a friend's shoulder to get him or her to join in; with open hand they beg for help from bystanders; they flee to a protector and, in safe proximity, shout and gesticulate at their opponent.

Supportive relationships were my principal research topic at Arnhem. Together with a changing team of students, I collected thousands of observations in the form "Individual A supports B against C." My earlier book, *Chimpanzee Politics* (1982), provides many details on the popularity factor in leadership campaigns, the tactic of isolating rivals from their rank and file, and the role of females in male power takeovers. If presidential candidates take a sudden interest in women, listen to their problems, and hug their children, there are parallels in chimpanzee males who groom females and play gently with infants, especially during periods of status struggle. Let me give a quick summary of power relationships, emphasizing sex differences.

Each group member has definite personal preferences when intervening in conflicts. The preferences of females and youngsters are stable, while those of adult males change over the years. The most powerful female coalition at Arnhem is the one between Mama and her friend Gorilla (a chimpanzee!). From the very beginning of the colony in 1971 Mama and Gorilla have fervently supported each other, against the most dangerous enemies. These females were acquainted even before that time; from 1959 on, they had lived together at Leipzig Zoo. Already there, wrote director W. Puschmann in a letter to me, they operated as a team against their cagemates. Almost all lasting female bonds in Arnhem are based on such common histories. Contrast this with Yeroen and Luit, two adult males, who also shared quarters in another zoo before coming to our colony; in Arnhem they had a number of different allies over the years, and they never established a stable friendship between themselves.

Yeroen ruled the colony for three years. When his position was threatened by a coalition between Luit and Nikkie, he received massive female protection. This did not save Yeroen, however, and in the fall of 1976 Luit became the so-called *alpha male*, or top-ranking male. Soon Nikkie turned from being Luit's supporter to being his main rival. The two of them competed every day over contact with the fallen leader, Yeroen. Both tried to sit and groom with Yeroen, and prevent the other from doing so. Luit lost this competition, which lasted about a year, because of Yeroen's growing preference for Nikkie. In 1977, with the old leader's backing, Nikkie was able to challenge Luit and reach the top. As soon as this power takeover was behind them, however, Yeroen tried—again with female support—to dethrone Nikkie. He was not successful, because his protracted battles with Nikkie played into Luit's hands. Luit could not be dominated by either of the other two males alone. Rather than endangering the new coalition any further, Yeroen settled for an influential position as Nikkie's second.

Generally speaking, males who are rivals one year may be allies the next, and vice versa. To understand this flexibility we have to distinguish between *coalitions*, as expressed in mutual support between two individuals, and *social bonds*, as expressed in affiliative behavior such as sitting together and grooming. If we assume coalitions to be part of social bonds, a flexible support network can come about only by changes in the bonding partners. Our data demonstrate that such changes do not occur; the social bonds of males are fairly stable. Instead, we found that coalitions simply do not depend on male bonding. Whereas females spring into action mostly to defend their offspring or closest friends, male coalitions are much harder to predict, as males frequently team up against individuals whom they normally prefer as grooming and contact partners.

Male coalitions are instruments to achieve and maintain high status. There is little room for sympathy or antipathy in such opportunistic strategy. The dissociation between a male's affiliative preferences and his coalitions is most marked in periods when he is jockeying for position. Adult male chimpanzees seem to live in a hierarchical world with replaceable coalition partners and a single permanent goal: power. Adult females, in contrast, live in a horizontal world of social connections. Their coalitions are committed to particular individuals, whose security is their goal. Some females, such as Mama, do exert considerable power in the group, but never at the cost of their relatives or best friends. Not once in the course of my research have I seen Mama turn against her friend Gorilla.

We know from psychological experiments with human subjects that in Western cultures men and women show similar differences. For example, people may be placed in a competitive situation, usually a game, which they can win only through cooperation with others. Men, when forming coalitions, are sensitive to the power distribution among the players and to strategic considerations, whereas women select partners mainly on the basis of personal attraction. Since attractiveness is more stable than strategic value, men are characterized by maneuverability. This trait is also visible in politics. Tancredo Neves, president-elect of Brazil, neatly summed up the male attitude in this arena: "I have never made a friend from whom I could not separate and I have never made an enemy that I could not approach."

To determine how widespread this sex difference is, and the

circumstances under which it appears, we need research on humans in a great variety of cultures. We also need observations on feral chimpanzees. Our present knowledge of chimpanzees in the wild tends to support the above picture. We owe this knowledge to two admirable ongoing field projects, both in Tanzania. One was started in 1960 in Gombe National Park by Jane Goodall, the other in 1965 in the Mahale Mountains by Toshisada Nishida and other Japanese primatologists. Goodall witnessed several power takeovers in her chimpanzee community and has repeatedly emphasized the importance of male coalitions. Nishida reports how an old male in the Mahale group regularly changed sides between two younger males who each needed the old male's support in order to dominate the other. Nishida, who was able to keep track of the fighting males in the jungle for several months, speaks of "allegiance fickleness." In this way the old male created a key role for himself, one that paid off in sexual privileges.

This is the same tactic used at Arnhem by Yeroen during the first year of Nikkie's reign. Yeroen could count on Luit's help to chase Nikkie away from sexually attractive females, and on Nikkie's help to chase Luit away. By skillfully playing off the two younger males against each other, he enjoyed the largest number of sexual contacts in the colony. Such configurations demand a distinction between *formal rank* and *power.* Formal rank is expressed in ritual encounters with impressive hair raising by the dominant and with greeting grunts and bowing by the subordinate. Both Nishida's old male and Yeroen made up for their lack of formal dominance over younger and stronger males by significant manipulative power.

The male hierarchy is heavily formalized, that is, males frequently communicate their status to one another. Among such fierce competitors, formalization is a requirement for relaxed relationships. Serious fighting erupts when status communication breaks down, and the winner applies the mechanism of conditional reassurance to reinstate it. The formal hierarchy may be seen as a device to maintain cohesion in spite of rivalry. Thus, although the Arnhem males have twenty times as many aggressive incidents among themselves as the females, they associate and groom at least as much as the females do. By comparison, the female hierarchy is rather vague. Since status communication is rare among females, it is difficult and almost useless to assign them positions on a vertical scale. The same is true of feral chimpanzee females.

The high frequency of male reconciliations—even taking into account the increased number of male conflicts—may be related to these sex differences. In the first place, a clear-cut hierarchy provides a ritual format for reunions after disagreement. Reconciliations among males are often preceded by a confirmation of formal status. For example, the dominant stands upright with all his hair erect and in one mighty gesture passes a raised arm over his ducking partner before they proceed to kissing and grooming. Second, the unreliable, Machiavellian nature of the male power games implies that every friend is a potential foe, and vice versa. Males have good reason to restore disturbed relations; no male ever knows when he may need his strongest rival. Holding grudges may cause isolation, which within the coalition system amounts to political suicide. In human politics as well, success demands an ability to compromise, forgive, and forget. In view of the statement quoted earlier, it is not surprising that Tancredo Neves was known in his country as the Great Conciliator.

For female chimpanzees the situation is entirely different. Their coalitions withstand time, overlap personal preferences and kinship bonds, and are relatively unimportant in status contests, as females are much less dominance oriented. For them it is of paramount importance to keep good relationships with a small circle of family and friends, but there is little reason to make up after fights with others. Over the years I have gained the impression that each female in the Arnhem colony has one or two absolute enemies, with whom reconciliation is simply out of the question. Rather than calling females less conciliatory than males, I prefer to call them more selective. The distinction between friend and foe seems infinitely sharper for females.

Bonding and solidarity are stronger among the Arnhem females than among their free-living counterparts. This is because
competition over food is not a life-or-death matter in a zoo. In
the wild, where food can be in short supply, female chimpanzees avoid competition by living dispersed throughout the forest, each accompanied by her offspring of up to ten years of
age. This rather solitary life explains why the social mechanisms
that allow males to keep tensions under control are less developed among females. Adult male chimpanzees have a very real
need to cope with competition, as they often travel together in
bands. In addition to each individual's personal motivation to
stay in touch with the male core, it is of vital interest for all the
males of a community to close ranks in the face of aggression
from males in neighboring territories.

Female chimpanzees maintain intimate relationships with a small circle
of friends and kin. Here an adult female holds her daughter still while
grooming her face. (Yerkes Primate Center)

Not surprisingly, in view of this male-female difference in life-style, Jane Goodall in *The Chimpanzees of Gombe* confirms the sex difference observed in the Arnhem colony. In wild chimpanzees as well, attacks by males are more often followed by reassurance than are attacks by females.

Do similar sex differences exist in our own species? A touchier issue can hardly be raised. Opinions are divided even among feminists. For example, Marilyn French in *Beyond Power* classifies hierarchical structures in human society as typically male, and egalitarian networks as typically female, a distinction resembling the one found in chimpanzees. But the author also depicts women as the most peaceful creatures on earth, devoid of competitive tendencies. She believes that prehistoric people lived in societies ruled by women: "The matricentric world was one of sharing, of community bound by friendship and love, of emotional centeredness in home and people, all of which led to happiness." Apart from the lack of evidence that such a dream-world ever existed, I dare to question even its theoretical existence.

Do all women love and support one another? A few selected friends, yes—just as chimpanzee females—but in general? And are women always accommodating among themselves? A Dutch swimming coach, Marianne Oudkerk-Heemskerk, explained in an interview with *NRC-Handelsblad* (March 5, 1981) why she prefers to work with boys. During her fifteen-year career the trainer has had many more problems with jealousy and grudges among girls than with the more straightforward competition among boys: "I would rather see boys, so to speak, punch each other in the face when they disagree and drink a beer together one hour afterward than girls who may maintain a particular discord for months."

Because the stereotype is hackneyed, it is not necessarily untrue. We need systematic studies on relationships within convents, nurses' homes, women's colleges, and so on, to verify whether Woman the Dove is myth or reality. Females compete in many subtle ways. A female chimpanzee may, for example, instigate an attack by a male friend on another female. She will sit next to the male, arm around his shoulder, directing a few high-pitched barks at her rival. When the male obliges by charging at the other female, we may score it as yet another instance of male aggression; but that is only because we use such crude measures. Among humans as well, female competition may often be overlooked.

From a biological perspective, absence of aggression among females simply does not make sense. Resources are limited, and each individual, male or female, tries to survive and reproduce. Feminist primatologists have therefore begun to challenge the notion of one quarrelsome sex and one peaceful sex. New data reviewed by Sarah Hrdy in *The Woman That Never Evolved* demonstrate that female monkeys compete just as intensely as males do, albeit for different reasons; males chiefly compete over mates, females over food for themselves and their progeny.

The only quibble that I have with Hrdy's views is that she describes female primates as strongly dominance oriented. This may be true for many monkey species—such as baboons and macaques—in which females do form definitive hierarchies, but it does not apply to our closest relatives, the apes. Status rituals are very rare among female chimpanzees; between some of the Arnhem females I have not witnessed any such ritual in six years. Also, whereas male chimpanzees engage in fierce dominance struggles, females do not. Actually, the portion of male aggression that goes into rank-related affairs is so large that one could argue that their social structure creates rather than reduces aggression. The crucial point is that this structure, and the mutual reassurance between dominants and subordinates that it involves, makes rivalry among males *less divisive* than among females. The males' hierarchy canalizes aggression in predictable directions and unifies the competitors.

Yet this thesis should not be taken to mean that males are about as peaceful as females. That would be a gross distortion of reality. In both chimpanzees and humans, males are more inclined to physical violence. They are often the troublemakers.

What I am saying is that we should not assume that females do not compete *at all*. The most interesting difference between the sexes is, I feel, not in the amount of competition but in the form it takes and the effect it has on relationships. In fact, the erroneous image of women as noncompetitive may result from a tendency similar to that of female chimpanzees, that is, a proclivity to stay away from rivals. As Lillian Rubin observed, "So, rather than acknowledge our competitive feelings—yes, even the wish to best one another sometimes—we distance ourselves from the object of competition, thereby damaging the very closeness we wish so much to protect."

A Coalition Breaks

When peaceful solutions fail or are ignored, violence breaks out. This danger is not always obvious; aggression can be buffered and cushioned so well that the resulting peace is taken for granted. When the Arnhem males had an intact hierarchy, violence was virtually unknown. They carefully maintained their social structure by means of status confirmations, calming gestures, and grooming. Even occasional rank reversals were achieved without life-threatening fights. Beneath the surface enormous tension could be felt, especially when males hovered near a sexually attractive female, readily recognizable by her pink genital swelling. Yet things remained under control. Since this was the situation during the first nine years, we came to regard it as the normal state of affairs. After so many years of relative calm we were ill prepared for the temporary breakdown of the system in 1980. For the following account it is necessary to go back to the formation three years earlier of the ruling coalition between Yeroen and Nikkie.

Paradoxically, Luit was defeated by the other two males in 1977 because he seemed capable of standing on his own legs. Within months of achieving alpha status he had gained broad female support. In order to keep his rival, Nikkie, under control he required only that the old leader, Yeroen, be neutral. Nikkie, in contrast, needed Yeroen's active support to have any chance at achieving top status. Yeroen's eventual decision to join Nikkie is understandable. Being the right-hand man of a new leader who depends on you completely gives you more clout than joining a more powerful ally who will inevitably try to monopolize the privileges of high rank. Luit's fate is illustrative of the rule "Strength Is Weakness," known from human coalition theory; strong parties seem almost to *invite* cooperation against themselves.

Although Nikkie was made the formal alpha male, submissively greeted by each and every member of the community, his first year in that position was weak. As described earlier, he could not prevent his partner, Yeroen, from being the most successful at mating. Females paid their respects to Yeroen

The coalition of Yeroen (*left*) and Nikkie dominated the Arnhem colony for more than three years. The younger and stronger Nikkie was the formal head, but he was totally dependent on Yeroen. (Arnhem Zoo)

more often than to Nikkie; their submission to Nikkie was not spontaneous. His position was considerably strengthened, however, when both he and Luit suddenly ceased to let their sexual jealousy be exploited by Yeroen. Overnight the old male's game was up. He would still occasionally scream and hold out his hand to one male if the other approached a female in heat, but they both ignored such requests. As a result, most matings in the second year were by Luit and Nikkie. Subsequently, their "nonintervention treaty" developed into a sort of coalition. They would regularly charge over the island as a team, displaying together at Yeroen if he sat for too long, too close to an attractive female.

This was the situation in 1979, at the time I wrote *Chimpanzee Politics*. The balance in the male triangle, which had first been determined by Yeroen from below, was now determined by Nikkie from above. His cooperation with Luit was strictly limited to the sexual context, and he remained dependent on Yeroen to stay dominant over Luit. A crucial element of Nikkie's strategy was to prevent contact between the other two males. One of my students, Otto Adang, recorded so-called *separating interventions*, in which one individual breaks up a contact between two others, usually by displaying right in front of them, throwing sticks and stones, or threatening to charge. The majority of such interventions were carried out by Nikkie. He aimed particularly at Luit's contacts with Yeroen, but also at those with high-ranking females. Luit obviously knew the rules; Nikkie needed only to stare in his direction and Luit would scratch his head, look at the sky, and silently stroll away from a grooming partner. Yeroen often acted as Nikkie's assistant, and sometimes as the instigator of these interventions. Loudly hooting, Yeroen would draw Nikkie's attention to a contact by Luit, and then the two of them, charging shoulder to shoulder, would scatter the group in which Luit was sitting.

The serious tensions that erupted in 1980 in the Arnhem colony were caused, I feel, by Yeroen's growing frustration. He had brought Nikkie to power and continued to help him keep Luit socially isolated. But as soon as sex was at stake, Luit was allowed to undergo a metamorphosis, from submissive to self-confident, because Nikkie treated him very tolerantly. These situations did not occur too often, because only females without suckling offspring become receptive, and then for only about fourteen days of their monthly cycle. Still, Yeroen's gains from backing Nikkie seemed to be dwindling.

After countless minor incidents, in which matings by Luit led to tension between the group's two rulers, Yeroen and Nikkie, the first real attack was observed on July Fourth. A female named Krom sported the characteristic pink swelling that attracts chimpanzee males. In the morning we observed rapprochement between Luit and Nikkie. When Yeroen sexually invited Krom, the other two males approached him with hair on end. Yeroen left the female, but pushed Nikkie and hit Luit. Then all three males screamed and Nikkie and Luit briefly mounted each other.

Hours afterward, all three males were sitting under a tree with Krom above them. When Luit started to climb up to her, Yeroen yelped and looked from Luit to Nikkie. Luit hastily returned to the ground and approached the others. The three males hooted in chorus. After a few minutes, however, Luit went back into the tree. Now Yeroen burst into loud screaming directed against Luit, while holding out his hand to Nikkie, begging him for support. Nikkie walked away from the scene. Yeroen responded by making a highly unusual surprise attack on Nikkie, jumping on him from behind and biting him in the back. He seemed furious that Nikkie had ignored his request to stop Luit.

Two days after the above incident an unobserved fight took place in the night cage of the three males. In view of the injuries found on two of the males, this must have been the most intense fighting since the establishment of the Arnhem colony. Nikkie showed deep injuries on the tips of many fingers and toes, as well as on his bottom and ear. Yeroen's fingers and toes were bitten and swollen; he was missing several nails and the tip of one toe. Luit, in contrast, showed only one superficial scratch. Male fighting does not usually result in injuries, but if it does the wounds are almost invariably found on hands and feet. It was not the location of the injuries, therefore, but their number that made this fight exceptional. In addition, none of our males had ever before lost part of a finger or toe.

Although no winner or loser could be determined on the basis of a mere injury count, Nikkie clearly behaved as if he were the loser. Until that night he had been a very impressive, big alpha male; now he looked unrecognizably small, depressed, and pitiful. Even though it did not seem as if Luit had been much involved in the physical battle, he emerged as the new dominant male. While this is hard to understand on a one-on-one basis, it is obvious when we consider the male triangle as a whole. Previous years had provided plenty of evidence that Yeroen and Nikkie needed each other's support, and that Luit would regain control as soon as their coalition collapsed. Luit was the first male to become alpha overnight, apparently without having had to fight for the position. My interpretation is that the breakup between Yeroen and Nikkie had created a power vacuum, which was immediately filled by Luit. He became alpha *by default*.

The alternative reconstruction of the incident is that Luit had single-handedly injured and defeated both other males. But, as confirmed by subsequent events, such a feat must be considered beyond Luit's physical capabilities.

Deadly Violence

The four adult males had been kept together in two interconnected night cages since May 1978. The youngest male, Dandy, more or less had his choice of sleeping arrangements. During some periods he slept with the other males, and during other periods he preferred to sleep apart (as he did during the entire 1980 episode described here). After the fight between Yeroen and Nikkie we decided to keep the three senior males out of the group for one week and put them together only when supervision was available. All went well; the three males spent the

A status ritual demonstrating that Luit (*right*) is the new boss. While Nikkie bows, Luit pushes himself up and raises his hair. The result (*next photo*) seems to be a striking disparity between the two males, whereas in reality they are about equal in size. (Arnhem Zoo)

days in one of the large indoor halls and the nights in separate cages. After one week they were reintroduced into the group At night we still kept them apart.

In the course of time, however, it became increasingly difficult for caretakers Jacky Hommes and Loes Offermans to separate the males at night. Yeroen always tried to enter a cage with Nikkie. If he succeeded, Luit became very upset, refused to go into his own cage, and occasionally even attacked the caretakers through the bars. If, on the other hand, Luit and Nikkie happened to enter a cage together, Yeroen showed the same response. It seemed that neither Luit nor Yeroen wanted to be left out if the other two males managed to get together.

After approximately seven weeks we decided to leave the matter up to the males themselves. If they strongly wanted to

sleep together, they were allowed to do so. Otherwise they were kept apart. This decision relieved the caretakers of the stressful and time-consuming job of isolating the males, with only trapdoors and a water hose as instruments—a task that sometimes took until late in the evening. At the time I held the philosophy that chimpanzees are better at estimating the possibilities within their own relationships than we human observers are. Perhaps this assumption is still correct; the dramatic consequences of the males' desire to spend the nights together, and of our decision to let them, does not necessarily mean that they were unaware of the dangers.

During the entire period between the first serious fight and the second, Nikkie behaved extremely submissively toward Luit, sometimes literally groveling in the dust for him. Yeroen showed much less submissiveness and often counterdisplayed if Luit approached him with hair on end and in a dominant posture. But even Yeroen's rare submissive grunts to Luit were a big change as compared to their previous relationship, which had been balanced by Nikkie's protection of Yeroen.

The first days after the males' reintroduction into the colony were marked by massive, intensive grooming. All the chimpanzees gathered in ever-changing grooming clusters. Much attention was given to Nikkie's injuries (more than to Yeroen's), although Luit often ousted females who stayed too long with Nikkie. As far as the group as a whole was concerned, Luit's leadership was a significant improvement. Remarkable peace and playfulness reigned, even among the older females, who normally never gallop around uttering the throaty chimpanzee laugh. Luit took the so-called control role, acting as arbitrator in disputes with great authority and impartiality. It must have been his alertness to tensions in the group, and his habit of standing impressively between screaming parties until they calmed down, that created the same state of harmony as we had observed during his earlier period of leadership, in 1976–1977.

The peace did not extend to relations among the adult males, however. There remained signs of tension and instability. It is difficult to summarize the changing situation. One day we might note in our logbook that Luit seemed to be forming a coalition with Dandy (who had suddenly become more active); the next day we would see Luit and Nikkie make a joint intimidation display in large circles around the other two males, who screamed in panic, and we would predict a future Luit-Nikkie coalition. The males seemed to be trying out all possible supportive combinations. Except, that is, for the Yeroen-Luit combination. Although Luit appeared very self-confident, there were subtle indicators of his wariness of Yeroen. If the old male came to sit not far from him, Luit would look uneasy for a few minutes before leaving the scene. On several occasions I was close enough to hear Luit sigh deeply once he sat down again at a distance from Yeroen. He also sometimes sat huddled up, with his arms in a knot or squeezed between his knees, in a sort of fetal position not at all characteristic of him.

The options for Yeroen seemed the same as the previous time when Luit ruled the colony. Joining Luit would have little advantage, and Dandy seemed too young to pose a serious threat to Luit, so all that remained for Yeroen was to restore the broken coalition with Nikkie. Our standard measures of coalition include the tendency of two individuals to intervene in each other's conflicts, the direction of these interventions (for or against), the number of joint displays, and so on. By all measures the Nikkie-Yeroen coalition was considerably weaker than before. Yeroen did his best to get the relationship back to normal. He would scream in frustration and follow Luit and Nikkie around whenever they walked together. When Luit was gone, he himself would try to walk, sit, or groom with Nikkie. His tactics were not always successful, for two reasons. Without exception, Luit began a noisy intimidation display if he noticed contact between the other two males, and he usually managed to separate them. Second, Nikkie often avoided contact with Yeroen by himself, even if Luit was out of sight; he gave the impression of disliking Yeroen and mistrusting his intentions. With the healing of his wounds, however, Nikkie's negative attitude gradually disappeared.

During the night of September 12 to 13 the males' night cages turned red with blood. When we arrived in the morning, the males had apparently already reconciled; they were relatively calm, and Jacky had trouble separating them. Luit made strong efforts to stay with the two other males, which was quite remarkable in view of what they had done to him. The fact that he did so demonstrates the male chimpanzees' deep need to belong—a need that makes sense in view of their natural life, in which lone males probably cannot survive intercommunity hostilities, as I will explain shortly.

Luit had many deep gashes on his head, flanks, back, around the anus, and in the scrotum. His feet in particular were badly injured (from one foot a toe was missing, from the other foot several toes). He also had sustained bites in his hands (several nails were missing). The most gruesome discovery was that he had lost both testicles. All the missing body parts were later found on the cage floor. Closer inspection, on the operating table, of Luit's scrotal sac revealed that, contrary to our expectation, it had *not* been ripped wide open. Instead, there were a number of relatively small holes. It was unclear how the testicles had come out.

For three and a half hours the zoo's veterinarian, Piet de Jong, and his assistant worked to save Luit's life. They cleaned his wounds and must have inserted between one hundred and two hundred stitches. In the evening, however, Luit died in his night cage, still partially under narcosis. The principal cause of death may have been the combination of stress and loss of blood. By the time of death the rest of the colony had entered the sleeping quarters. They were completely silent during the time that Luit's body was lying in his cage. The following morning, even at feeding time, hardly any sounds were heard. Vocal activity resumed only after the corpse had been carried out of the building.

During his brief period of dominance Luit acted with authority and self-confidence. Occasionally, however, the tensions with Yeroen would work on his nerves and he would sit, all huddled up, in a fetal position (*above*) very different from his typical regal posture. (Arnhem Zoo)

For a reconstruction of the second night-cage fight, it is important to know that Luit was the only male who sustained serious injuries. Nikkie did not show any damage at all, whereas Yeroen had only small scratches and cuts (his injuries, though large in number, were superficial). Since Luit was a strong male, definitely stronger than Yeroen and of at least the same strength as Nikkie, the unequal outcome of the fight is, in my opinion, explainable only by assuming a high degree of collaboration between Nikkie and Yeroen.

An alternative explanation, suggested to me during presentation of the case to a group of veterinarians, is that the other two males had mounted a surprise attack on Luit while he was asleep. A heavy blow or bite in the testicles might have paralyzed him for a moment, during which it would have been easy to attack him further. The question is whether such a paralysis through pain would have lasted long enough. Blood was smeared and spattered all over the floors, walls, bars, and even the wire roof of both night cages, and the straw was spread about in a disorganized way, suggesting a protracted struggle with a lot of chasing and escape attempts. Seeing the damage done to Luit and the mess in the cage, my guess is that the battle must have lasted for more than fifteen minutes and that Luit must have been far from immobilized.

On the morning of September 13, after Luit had been isolated for treatment, we released Nikkie and Yeroen into the group. Immediately there occurred an unusually fierce attack on Nikkie by a high-ranking female, Puist. She was so persistently aggressive that Nikkie fled into a tree. On her own, Puist kept Nikkie there for at least ten minutes by screaming and charging each time he tried to come down. Puist had always been Luit's main ally among the females. She must have followed the fight, for her night cage offers a view into the males' pens. Later in the day the group showed high interest in the two males, grooming and inspecting them.

From this day on, Dandy played a much more important role than ever before. He repeatedly sought contact with Yeroen, resisting separation attempts by Nikkie. In the course of subsequent months the new male triangle calmed down. On October 14 Yeroen uttered his first submissive greeting grunts to Nikkie since their fight during the night of July 6. In the following weeks they groomed each other frantically, ironing out, it seemed, the remaining tensions. The relationship between Yeroen and Nikkie became as close as the one they had had before, with up-and-coming Dandy now in the position of their common rival.

Reflections on the Dark Side

As I biked to the zoo that fateful Saturday morning, my thoughts and feelings were confused and hopelessly unscientific. Jacky's hasty description over the phone of the state in which she had found Luit still echoed in my ears, leaving me filled with sadness and disappointment, together with an impulsive verdict: Yeroen was to blame. He was, and still is, the one who decides everything in the chimpanzee colony. Nikkie, ten years younger, seemed only a pawn in Yeroen's games. I found myself fighting this moral judgment, but to this day I cannot look at Yeroen without seeing a murderer. Such sentiments should not be confused with facts, however. Nikkie must have been involved in the fight as much as Yeroen. Also, "murderer" implies an intent to kill, something impossible to prove or disprove in this instance.

My emotional state did not improve when I saw Luit sitting in a puddle of blood. Normally aloof toward humans, including familiar ones, he now sought contact and let me groom his head. "We shouldn't have let them be together!" the keepers and I kept repeating. But nothing in the history of the colony had prepared us for this drama. With Luit's death the Arnhem project entered a new phase. Some romantic notions were left behind, a change that occurred at a time when field-workers too were discovering a darker side to ape nature.

Observations in the wild of hunting and predation on baboons, colobus monkeys, bushpigs, duikers, and other forest

animals had already undermined the image of chimpanzees as our cute, amiable cousins. The witnessing of bloody fights between males of different communities, and of occasional cannibalism of infant chimpanzees by both males and females, was the final blow. These discoveries made the chimpanzee even more humanlike than previously thought possible, and for some scientists the species was now coming too close for comfort. Ashley Montagu, for instance, tried to escape the inescapable by ingeniously suggesting in a letter to the *New York Times* (May 2, 1978) that "under certain conditions chimpanzees behave more like humans than humans behave like chimpanzees."

Jane Goodall's accounts are especially relevant here, because they give a clue to the way Yeroen and Nikkie may have collaborated against Luit. After fission of the large Gombe community, adult males of the main group repeatedly invaded the smaller group's territory and managed to take it over by killing at least three of their males (two others disappeared in the same period and are suspected to have been assaulted as well). Each of the three observed attacks began as a swift, silent approach through the undergrowth by several males who had spotted a neighboring male alone in the distance. A coordinated attack of extended duration and extreme brutality followed. Pinned to the ground by one aggressor, the helpless victim was pounded, hit, trampled, and bitten by the others. They then left the victim behind in a state of shock, covered with deep wounds. Two of the victims were never seen again, and one only once, months later, in an emaciated state. Interestingly, Goodall's description of this encounter mentions that "his scrotum had shrunk to about one-fifth normal size."

Adolescent and young adult females are allowed to move across territorial borders, especially when they are in a sexually attractive state. Anne Pusey has speculated that the bright pink swellings of female chimpanzees serve as a long-distance signal

Yeroen. (Arnhem Zoo)

to potentially hostile males—much as if they are waving their passport. Older females with dependent offspring, on the other hand, have to be just as careful as roving males. The strength of chimpanzee xenophobia is evident in a telling incident of a confrontation between adult males and a stranger female near the periphery of the males' range. Goodall relates that the female responded to the threat by making submissive sounds and reaching out to gently touch one of the males. The male did not want contact, however. He instantly moved away from her gesture, picked a handful of leaves, and vigorously scrubbed his fur where she had touched him. Subsequently the female was surrounded and attacked; her infant was seized and killed.

Infanticide has also been observed at the other major study site, in the Mahale Mountains; but Japanese scientists have not yet witnessed deadly wars between males of different communities. Severe fighting does occur, and there are signs that it may lead to death: healthy males of one community disappeared one by one over the years until their territory was eventually taken over by two other communities.

In view of their extreme territoriality, male chimpanzees may almost be regarded as captives in their own group; they cannot leave their home range without running into great trouble. A feral male suddenly finding himself, like Luit, in a very uncomfortable position can only move to the periphery of the community range. In this no-man's-land he will have to keep one eye on his groupmates and the other on the boundary patrols of his neighbors. In both Gombe and the Mahale Mountains such outcasts have been observed. Their fate has been characterized as "going into exile." When social tensions subside, these males can usually return to the core area, but in at least one case this may have been an unwise move. A Mahale outcast was observed to return to a group in which the oldest male was reputed to be fickle in his allegiances. Soon after the outcast's return, the elderly male assisted him in reaching alpha status. A few months later, however, the new alpha male vanished—for unknown reasons.

If the mysterious disappearances of feral male chimpanzees

reflect death due to aggression within or between communities, the incident at Arnhem Zoo is not as unique as we initially thought. Of course, the captive condition cannot be ruled out as a cause, but neither does it seem sufficient as an explanation. Simple concepts, such as stress and crowding, fail to explain how the same males could have lived together under the same circumstances for nearly eight hundred nights without the occurrence of anything dramatic. Also, they were in no way obliged to spend the nights together. As I see it, the housing conditions provided an *opportunity*. They made the attack possible, but they do not explain its deeper causes.

The first night-cage fight seemed related to a growing discrepancy between Nikkie's dependency on Yeroen and his increased leaning toward Luit in the sexual context. By bringing Nikkie to alpha rank, Yeroen had regained both respect from the females and a good share of sexual activity. I tend to interpret this as a "deal," the fulfillment of which was closely monitored by the old schemer. When Nikkie failed to keep up his end of the deal, Yeroen ended the cooperation. Suddenly all three males were in an awkward position. Nikkie lost his top status; Yeroen seemed to have few options other than trying to mend the coalition; and Luit became alpha, but was insecure in Yeroen's presence. There is no way of knowing whether the second big fight was a purposeful attempt to solve these problems by eliminating a rival, an act of blind frustration by Yeroen and Nikkie, or something else. The fact is, though, that it did resolve the tensions.

Even if Luit had survived, he would presumably have assumed a different role in the group because of his castration. This is a highly unusual type of injury. The extensive literature on primates mentions only very few other examples, such as the Gombe male with the reduced scrotum and a rhesus monkey in India who received a deep laceration in a testicle during an attack by four other males. I myself once witnessed a fight among male long-tailed macaques in which one male confronted an opponent face to face while a third male bit him from behind, resulting in the loss of one testicle. Further, field-workers have told me of two independent occasions on which immobilization of a feral male baboon led to an immediate assault by nearby males on the slightly drowsy subject. In both cases the aggressors were chased off by investigators, but not before their victim had sustained deep groin injuries close to the scrotum.

I have been unable to confirm several other verbal communications about scrotal injuries. These reports concerned captive primates, including chimpanzees. In this time of animal rights activism, laboratories and zoos have little interest in bringing such information into the open. Although I share many of the movement's ideals, its stifling effect on the information flow is regrettable. In order to improve animal management we need to know about both successes and failures. Fortunately, the director of Arnhem Zoo, Anton van Hooff, fully recognizes this and has without hesitation approved the publication of details on Luit's death.

In humans, castration may serve a practical purpose. It has been performed to create high-voiced castrati, to cure rapists, and to exclude certain slaves (the so-called eunuchs) from procreation. Most commonly, however, mutilation of genitalia is an act of violence and suppression. Clitorectomy, for example, still practiced on millions of women, is the cruelest expression of male dominance. The operation kills the capacity for sexual pleasure, thereby increasing the control of fathers and husbands over daughters and wives, who feel less tempted to seek adventures outside the home. No one knows how the practice came into existence, but it might have started as a punishment for adulterous women. In much the same way, severed genitals stuffed in the mouth of a Mafia victim mean that the dead man had paid too much attention to a particular woman. That sexual jealousy should lead to acts against the critical organs is perhaps not surprising in the case of humans, but Luit's fate follows the same pattern. Whether this involved a conscious connection on the part of his attackers remains a mystery.

Another situation in which humans emasculate enemies is

during wars or coups d'état. A recent report comes from Surinam. Fifteen people, suspected of planning to overthrow the junta of this South American nation, were executed in 1982. According to an eyewitness, some of them had been castrated. As a constant reminder of this possibility, many languages are rich in intimidating expressions concerning the male sexual organs. Universally, men are preoccupied with and care about their "jewels." Public reference to them, in an aggressive tone, represents the ultimate menace, as in the following quotation from James Clavell's *Noble House,* hissed by a woman in gutter Cantonese:

"If I say the word he'll have those insignificant peanuts you call your balls crushed off your loathsome body an hour after you leave work tonight."
The waiter blanched. "Eh?"
"Hot tea! Bring me fornicating hot tea and if you spit in it I'll get my husband to put a knot in that straw you call your stalk!"
The waiter fled.

In contrast to Konrad Lorenz's picture, in *On Aggression,* of humans as the only mammals who kill members of their own species, biologists now view humans as relatively pacific. Death through intraspecific violence is not a daily occurrence in the animal kingdom, but it does happen both in the wild and in captivity, and not merely as the result of mishaps. In his discussion of these new insights, E. O. Wilson noted that murderous behavior often is discovered only when the observation time devoted to an animal species passes the thousand-hour mark. He surmised that one would have to spend many more hours of watching an average human population to see the same behavior.

Perhaps we are not talking about the same behavior at all. It is sometimes argued that human aggression is cultural, whereas animal aggression is instinctive. This is a false dichotomy. It is like trying to decide whether pit bull terriers (responsible for twenty-one of the twenty-nine deaths due to dog attacks in the

United States between 1983 and 1987) are so dangerous because of their inbred ferocity or because of the way their owners treat them. Obviously, both genes and training play a role; it is *easier* to turn a pit bull than a golden retriever into a killing machine. Similar reasoning applies to human aggression. Each child is born with the potential to develop aggressive behavior—and in some children this potential is probably stronger than in others—yet the precise outcome depends on the child's environment. So if ethologists claim that humans have an aggressive nature, they mean that members of our species learn aggressive behavior quite easily. This is not the same as saying that the occurrence of violence and war is beyond our control. There is plenty of latitude for culture to exert influence: both violence and nonviolence can be taught.

A series of experiments by Albert Bandura and his coworkers has demonstrated that children readily imitate aggression. Normally, when left alone in a room with a large doll and other objects, children will play quietly. If they have witnessed an adult kicking, hitting, and yelling at the same doll, however, the children will subsequently beat it violently. They will copy the model's verbal abuse and fighting techniques, add aggressive variants of their own, and subject other play objects to their rage.

There undoubtedly is a strong learning component to aggressive behavior. It applies to both humans and animals; imitation plays a role in aggression among chimpanzees as well. This became distressingly clear in the period following the fatal fight in Arnhem. I received the first hint when witnessing an attack by the adult female Tepel on Dandy, who had provoked her by repeatedly displaying near her and hitting her son, Wouter. When Tepel chased him, the high-pitched barks of other females must have encouraged her, because she suddenly increased her speed in an attempt to grab him. Twice she dived beneath Dandy, trying to bite him between the legs. Since Dandy is much faster than any female, Tepel did not succeed. Nevertheless, from the shrill sound of Dandy's screams and his panicky flight, it was evident that this fighting technique had

Tepel (*left*) jumps over her son to defend him against an attack by Dandy (*right*). This is the beginning of an incident that subsequently led Tepel to attempt (outside my camera range) to bite Dandy in the testicles. (Arnhem Zoo)

come as a total surprise. In many years of watching the colony I had never seen an attack even remotely resembling this one. Tepel, who shares her night cage with Puist, may have learned the below-the-belt technique from watching the assault on Luit. The incident with Dandy took place one month after Luit's death.

Several months later, one of the females had an unusual injury on her belly. Its size and shape indicated that it must have been caused by the sharp canine teeth of an adult male. The use of these dangerous weapons against females, exceptional in itself, had until then always been limited to less vulnerable body parts such as the back and shoulders. It is unlikely that this abdominal injury was a mere accident; male chimpanzees show incredible motor control, even during the wildest fights. In addition to the above two cases, Otto Adang, who succeeded me at Arnhem, has observed three fights in the outdoor enclosure in which males lost phalanges of fingers or toes. This type of lasting mutilation, also observed in a few feral chimpanzees, had not occurred in our colony prior to 1980. With the killing of Luit, we seem to have crossed a threshold toward a higher risk of damaging aggression.

Probably the apes themselves were aware of this dangerous change in the social atmosphere. One indication is that they intensified their peace efforts. Tine Griede compared observations from before Luit's death with those afterward. She had collected measurements of both the frequency of reconciliation and the amount of "hesitation" involved. Hesitation was determined by counting how many approaches former opponents made before establishing contact. Aborted or rejected attempts are quite common; one reconciliation may take several approaches. Tine found that in the months following the fatal fight adversaries not only made up more often but also in a less hesitant manner than previously. Perhaps the chimpanzees were just as shocked by the incident as their keepers and were determined not to let it happen again.

I left behind in Arnhem a relaxed, thriving colony of chimpanzees. There had been no births in 1980, the year of Luit's death, but by the end of the next year we had a baby boom: three new infants and three pregnancies. Two females who nobody had dared to hope would raise offspring did just that. Spin, who had several times rejected her newborns without even touching them, in 1981 accepted one. Puist, a masculine-looking female who had always refused to mate—and instead had mounted females herself—became pregnant as a result of Nikkie's perseverance. She turned out to be a perfect mother.

Self-Awareness and Chimpocentrism

The grisly stories in the preceding pages of violence among chimpanzees stem from twenty-five years of intensive research at Gombe National Park, twenty years in the Mahale Mountains, and fifteen years at Arnhem Zoo. Violence is not the normal state of this ape's social life. It is there as an undercurrent, a constant threat, but chimpanzees keep their head above the surface 99 percent of the time. And when aggression does escalate, I would hesitate to ascribe it to a failure of peacemaking efforts. Peace may not be the preferred option under all circumstances, and chimpanzees may deliberately choose the alternative.

In general, conflict management among these apes is extremely efficient. The stories of violence only suggest that chimpanzees have compelling reasons to balance relationships. Some of the very behaviors used for this purpose hint at the dangers of failing to reach peace. For example, male chimpanzees often finger each other's scrotum at moments of mild tension, a gesture irreverently known among field-workers as ball bouncing. Is there a more convincing way of indicating friendly intentions than by touching these vulnerable parts? The pattern is also known from a number of human cultures. New Guinean tribes, for example, have a greeting in which the scrotum is briefly touched with an upward movement. In 1977 Irenäus Eibl-Eibesfeldt documented this gesture in a publication that described the response when he gave a piece of sugarcane to a six-year-old boy: "He gave expression to his great enjoyment by pushing his right hand from behind between my legs and gently stroking my scrotum through my trousers three times. At the same time he gave me a bright smile."

Kissing is another intriguing behavior. Although its message is quite different from that of biting, the two patterns resemble each other (at least when we exclude kissing on the mouth). Between kissing and severe biting there is a whole continuum of inhibited biting, mock biting, playful gnawing, and love bites. I like the thought of the kiss's deriving from its antithesis by increased control over the emotions and the jaw muscles. This would represent the ultimate paradox in peacemaking, one where *les extrêmes se touchent*. As speculated by Eibl-Eibesfeldt, mouth-to-mouth kissing may have evolved separately, via the transfer of masticated food from mother to young. This feeding technique is sometimes seen among apes and is also known from a large number of human cultures, including that of the ancient Greeks.

Chimpanzees have a habit of putting their fingers or the back of one hand between the teeth of dominant group members. A friendly gesture, it is also a test of the dominant's state of arousal and often is used in ambiguous situations. I experienced it myself when performing psychological experiments with two juvenile chimpanzees at the University of Nijmegen. Each day I spent hours in a room with them, and occasionally their constant mischievousness would get on my nerves. They would notice the slightest irritation and hurry over to fill my mouth with their big hands. Of course I never bit, but in the Arnhem colony I have seen quite a few instances when fingers were not treated so gently during appeasement attempts. Young chimpanzees of three years or less, who may have lacked the experience to judge whether the gesture was safe or not, were almost always the victims of such bites.

The great strength of chimpanzee reassurance mechanisms has far-reaching consequences, giving their society an extra dimension. Primatologists generally regard an individual monkey as dominant if his groupmates make way when he approaches. This simple criterion does not work with chimpanzees. Big adult males may be seen avoiding youngsters and females, at least when food is around. Taking another's food by force is uncommon, and even the highest-ranking individuals do not claim everything for themselves. Both in the wild and in captivity they allow others to tear off pieces, or give in to their begging by dropping part of their food. Generosity seems almost obligatory, because subordinates can get very upset and throw a tantrum if their requests are ignored. In order to avoid such embarrassing scenes, possessors of food—whatever their status—often walk away when a particularly greedy groupmate approaches. Avoidance by dominants and the taking away of food by subordinates are exact opposites from the usual rules in monkey society.

As early as the 1930s Henry Nissen and Meredith Crawford performed an experiment with pairs of young chimpanzees in adjacent cages separated by bars; one had food, the other did not. Between some of the subjects they saw threats and intimidation behavior, but others shared food in the course of friendly interactions involving grooming, play wrestling, and gentle stroking of the other's hands, face, and genitalia. The most characteristic begging gesture was the open hand held out in front of the food possessor. In Arnhem this same gesture is frequently seen after conflicts, as an invitation to the adversary to make up. The similarity in gestures and the importance of soothing body contact demonstrates a connection between reconciliation and food sharing.

Another example of chimpanzees following a more complex scheme is their response to sexual competition. In most primate species, adult males avoid one another and reduce the amount of contact when an attractive female is present. The Arnhem

Chimpanzees share food. A bundle of branches given to the group at the Field Station of the Yerkes Primate Center near Atlanta is first claimed by the dominant male (*left*). Soon others approach to obtain a few branches as well.

males, in contrast, try to overcome these tensions by grooming one another; the males aggregate rather than disperse in the face of sexual rivalry. There are even some indications that a sort of trade is going on. After a long grooming session among the males, a subordinate male may invite the female and enjoy a copulation without interference by the others. These interactions give the impression that males obtain "permission" for an undisturbed mating by paying a price in grooming currency. The phenomenon has been dubbed *sexual bargaining*.

This possibility should not come as a surprise, because chimpanzees are famous for their tradesmanship. Experimental studies indicate that the ability comes without any specific training. Every zookeeper who happens to leave his broom in the baboon cage knows there is no way he can get it back without entering the cage. With chimpanzees it is simpler. Show them an apple, point or nod at the broom, and they understand the deal, handing the object back through the bars.

Before a species can move from the principle of rights and privileges associated with dominance to the principle of sharing and trading, it needs to bring competition under control. Subordinates need to be able to calm aggressive dominants to the point where they become tolerant. Dominants need to control utter selfishness in order to reap the benefits of an exchange system. Perhaps sexual bargaining represents one of the oldest forms of tit for tat, one in which a tolerant atmosphere is created through appeasing behavior. We are dealing here with complex issues, because obviously more is involved than good will. Animals also need the capacity to make calculations and predictions before it becomes beneficial to relinquish the short-term advantages of exclusive possession in favor of the long-term advantages of mutual cooperation.

This capacity is particularly highly developed in humans. Our primary inclination still is to keep things for ourselves, but we also form networks of sharing and trading relationships that leave everyone with a better deal than is obtainable without cooperation. In spite of what skeptics say about human nature, strong restraints on open, utterly selfish competition do exist, even in the most materialistic of societies. Universally, people learn to give and take. Our intricate societies would be unthinkable without this ability.

Even in certain monkey species the required intelligence can be observed. Studies by Craig Packer, Barbara Smuts, and Ronald Noë on free-ranging baboons show that older males form coalitions to displace younger and stronger males from females in heat. The cooperation is often reciprocal; winners repay their teammates by supporting them on later occasions. In contrast to coalitions among chimpanzees, which unequivocally determine who will be boss of the group, the collaboration among male baboons serves to keep bosses from claiming too much for themselves. Lasting effects on status are minimal, as evidenced by the fact that baboon troops are usually dominated by young adult males who operate without supporters, their status based on sheer physical strength and agility.

These ingredients of primate social life—coalitions, conflict resolution, social tolerance, and strategic thinking—seem tightly interwoven, each one stimulating the development of the rest. Intelligence helps cooperation; reconciliation increases tolerance; tolerance makes sharing possible; the ability to trade comes with a further increase in intelligence, and so forth. The whole set of capabilities must have coevolved in an unbroken line of small steps, each one paving the way for the next. Undoubtedly, chimpanzees have more such steps behind them than monkeys, but I do not see a gap between their abilities. The sharing and trading capacities of chimpanzees, which extend to material goods, seem a logical extension of the exchange of intangible benefits, such as social favors, known from monkey societies. I stress the continuity because it has recently been suggested that there is a fundamental difference between monkeys and apes. According to this theory, which derives from research on self-recognition, only apes and humans have conscious minds. Let us explore that thesis.

When confronted with a mirror for the very first time, all primates are deceived; they respond socially with either threats or friendly gestures and try to look behind the mirror. In the course of time, however, an important difference between apes and monkeys emerges. Most monkeys go on treating their

image as a companion or an enemy until their interest gradually wanes. Apes, in contrast, start using the looking glass to inspect body parts (teeth, buttocks) that they normally cannot see. They also amuse themselves by making strange faces at their reflection, or by decorating themselves (placing vegetables on their heads, for instance). They can get quite absorbed in these activities and maintain a lifelong interest in mirrors as tools or toys. Wolfgang Köhler was the first to describe this striking phenomenon, in 1925; more than four decades later, Gordon Gallup designed a brilliant experiment to put it to the test.

The experiment consisted in smearing odorless, nonirritating paint above the eyebrows of an anesthetized chimpanzee with previous mirror experience. Upon recovery, as soon as he saw his reflection, the ape started rubbing and inspecting the bright red spot. After touching it, guided by the mirror, he smelled his fingertips, thereby eliminating all remaining doubt; he would have had no reason to smell his *own* fingers if he had interpreted the mirror image as *another* individual touching the red spots. This experiment proved that chimpanzees have self-recognition, which demands a concept of "self" as distinguished from "other." After many repetitions with different primate species, the conclusion thus far is that other than humans, only chimpanzees and orangutans make the connection between their mirror images and themselves.

Gallup in 1982 took these findings one step further. After linking self-recognition to consciousness and introspection, he listed other so-called empirical markers of mind. These include empathy, attribution of intentions to others, deliberate deception—and reconciliation behavior. There is indeed growing evidence for each of these abilities in chimpanzees, but does this mean that our closest relatives, in the words of Gallup, "have

A young chimpanzee plays with his mirror image. He is staring at his reflection, occasionally disturbing it by splashing the water with his hand. (Arnhem Zoo)

entered a cognitive domain which sets them apart from most other primates"?

Let us first back up a bit: does the absence of self-recognition in a mirror imply mindlessness? Michael Fox has drawn attention to the fact that Siamese fighting fish and parakeets keep attacking, courting, and even feeding their mirror image to the point of physical exhaustion, whereas dogs, cats, and monkeys clearly lose interest after a while. According to Fox, these mammals recognize that the duality between themselves and the other is an illusion. Is this not a first sign of self-awareness?

Not all monkeys lose interest in mirrors. Aziut, a long-tailed macaque at the Identity Research Institute in India, plays with mirrors all the time. He has spontaneously learned to use them to see what humans or dogs are doing behind his back. He directs the mirrors in a very precise way and may turn his head to compare the physical reality to the image. Aziut also experiments with the moving reflection of his hand in a mirror while he seizes food, and he sometimes tries to keep two mirrors facing each other by holding one on the ground with his foot and the other above his head. These are not exactly the games of an animal fooled by mirrors.

Let me add the example of the adult rhesus monkey females who will be treated in the next chapter. Their pen has a row of six large, reflecting observation windows near the ceiling, more than 5 meters above the floor. Every birth season we see females place their newborn baby on the floor, walk a few steps, and intently stare up at one of the windows, shifting their head as if searching for a particular reflection. Then they pick up the baby again. They start doing this within a day or two of giving birth. All the windows are used for this activity, regardless of which one we are standing behind.

I cannot explain this behavior. Perhaps the mothers like to have a look at their infant from a distance of more than 10 meters, without the risk of leaving him too far behind. They never stare at the windows in this particular way when carrying their baby, or when another female's youngster is walking free. They seem to connect their own behavior (placing their offspring on the floor) to the mirror image. That they do not, like chimpanzees, use the mirror to look at their own reflection may be a matter of how much interest monkeys have in themselves compared to such attractive creatures as their new infants. Fox speculates that apes and humans may simply have reached a higher level of narcissism.

In the absence of mirrors, it is more difficult to get information on self-awareness. But Craig Packer noticed in the field that the exaggerated yawning display of male baboons, to show their impressive canines to the world, depends on the condition of the teeth. Regardless of their age, males with broken or worn canines yawn less than males with healthy, big teeth. With no other males in the vicinity, however, males with poor teeth yawn as much as the rest. Packer does not speculate on the psychology of this yawn suppression, but I wager it comes close to self-awareness.

In short, the differences in consciousness between chimpanzees and most other nonhuman primates seem to be gradual rather than radical. The growing tendency to place the impressive mental capacities of chimpanzees on a pedestal has been called a chimpocentric bias by Benjamin Beck, and it is just as misleading as anthropocentrism. In the last couple of decades chimpanzees have frustrated linguists, psychologists, anthropologists, and philosophers seeking simplistic definitions of human uniqueness. Just as humans and apes share many psychological and mental traits, so there exists a continuum between them and the rest of the primate order. This holds true for all possible traits, including reconciliation behavior. Rather than viewing reconciliation as a "marker of mind," present only in hominoids, I would expect to find it in any species that lives in cohesive groups, with long-term relationships worth repairing after disagreement. The only capacities strictly needed are individual recognition and a good memory; both are present in many social animals, from hyenas to elephants and from dolphins to zebras.

Against this background I set out to study the peace strategies of monkeys.

CHAPTER THREE

Rhesus Monkeys

A Florentine comment on whether it is better to be loved or to be feared: "I reply that one should like to be both one and the other; but since it is difficult to join them together, it is much safer to be feared than to be loved when one of the two must be lacking."

—Niccolò Machiavelli

Before the Indian government halted the flow of tens of thousands of rhesus monkeys to laboratories in the West, one of the last groups to be trapped and shipped to the United States arrived in 1972 at the Wisconsin Regional Primate Research Center in Madison. Since that time the monkeys have been on public display at our facility in the Vilas Park Zoo. The group is used for behavioral studies and breeding. There are many breeding colonies all over the world, because the rhesus monkey remains the most common laboratory primate. From this hardy species the first "man" in space was recruited. We all have our Rh blood factors, named after the species. And if it were not for research on these monkeys, humanity would still suffer the devastations of polio.

Matriarchs and Matrilines

The group in Madison came from Uttar Pradesh, a state in the north of India near the Himalayas, where 90 percent of the

monkeys dwell in human villages, cities, along the roadsides, and in Hindu temples. Rhesus monkeys have lived in close contact with humans for centuries, and it is hard to say what their "natural" habitat is. It seems fitting, therefore, to introduce the main features of their social organization by imagining an abandoned village taken over by monkeys.

This fictitious village has a single street with a row of houses numbered, say, 1 through 10. Each house is owned by a matriarch, an elderly female whose daughters have children of their own. The whole female lineage (daughters, granddaughters, great-granddaughters) stays in grandma's house. Sons have a marginal role. They start leaving the house at a young age, to play with male peers and to hang around the few big males of the village. When they reach adolescence, sons tend to leave the village altogether. Adult males, too, usually come from neighboring villages and have no connection to any of the female families. Male comings and goings depend on contests among the males themselves, and also perhaps on what the female community thinks of them. When resident males are challenged by males from outside the village, females have the option of protecting the status quo or supporting the newcomers. Successful males may strut up and down the street with their tails in the air, receiving a lot of respect from everyone, but the village is essentially a female domain.

The matriarch of house number 1 rules the female population with an iron fist. All of her daughters become dominant over females on the rest of the street. This process starts at an early age, so that it is not unusual to see a full-grown female being chased by a very tiny one. If the older female dares to fight back, the younger one screams loudly to mobilize her relatives. This mechanism works over the full length of the street. Thus, all the females of house 7 stand behind their offspring in confrontations with the inhabitants of houses 8, 9, and 10. The

Every rhesus monkey has a different face and a unique personality. This young adult female is named Thistle. (Wisconsin Primate Center)

result is a *status tradition* among the females, handed down from generation to generation. Some females are born with the equivalent of a silver spoon in their mouth, others are not.

What seems to hold the village together is the good connections between neighbors. Members of neighboring households spend more time together than females whose houses are farther apart. There are of course exceptions. Some top females have good friends living down the street, but on the average there is more contact among females of like status. Similarity in age is important as well; same-aged females prefer one another's company. This tendency is so conspicuous among the matriarchs that we speak of the "old-girl network."

This village analogy is dreadfully misleading in that it sounds almost logical for female monkeys to belong to a certain "house" and for "neighbors" to have a lot of contact. In reality, of course, there is no row of neatly numbered houses. What we see is a large number of monkeys continuously walking and running around. Somehow they have figured out the kinship group to which each individual, over several generations, belongs. They are also aware of the hierarchical position of their own kinship group in relation to the rest.

Matrilineal hierarchies were discovered in the 1950s by Shunzo Kawamura, Masao Kawai, and other Japanese scientists working on the red-faced macaques native to their country (popularly known as the "hear no evil, see no evil, speak no evil" monkeys). There are many different macaque species, the rhesus monkey being one of them. Matrilineal hierarchies are not restricted to the genus *Macaca*, however; they have also been described for female baboons and vervet monkeys. Note the tremendous difference from chimpanzee social organization. Female chimpanzees do not organize themselves in cohesive groups with a clear-cut hierarchy. The chimpanzee has a patriarchal system, with males forming the stable core and adolescent females moving away. Rhesus monkeys, in contrast, are what is called a female-bonded species. Young females stay with their mother and sisters to integrate for life into one of the tightest and most complicated social systems known in the animal kingdom.*

The Transfer of Rank

Upon arrival in Madison I immediately started identifying the monkeys at the Primate Center. This task is comparable to a schoolteacher's learning a new class of children. The only difference is that I had to invent all the names myself. Each feral-born female received a name starting with a different letter. These females are the founders of the matrilines; they range in age from fifteen to perhaps thirty years. The offspring of each matriarch were given names beginning with the same letter. Take for instance the number 1 lineage: it is headed by Orange, named after her bright fur color, and includes her daughters Ommie and Orkid and her grandchildren Oona, Ochre, and Oyster. We call them the O-family, or simply the O's. But remember that the word "family" refers to a kinship unit of adult females plus their dependent offspring; it has little to do with the human nuclear family. Unlike a starting field-worker, I had the enormous advantage that each monkey had been marked with a number on its chest—and that detailed records had been kept, which allowed me to know all the blood relationships for three generations back.

It is clear that in our Madison group, rank positions of daughters depend on those of their mothers. We can predict with almost total certainty which position in the hierarchy a newborn female will occupy when she moves into adulthood. Genetics does not seem to play much of a role in the transmission of

---

*Awareness of the intricacy of the social life of monkeys has changed our view of the research institution practice of keeping these animals in individual cages. This is now widely recognized as an ethical problem. Although physical exercise, increased space, and toys are all very well, it is clear that nothing matches contact with conspecifics when it comes to improving the well-being of laboratory primates. Advocacy of social housing is reinforced by the fact that it reduces maintenance costs even as it increases breeding success.

rank. It is a well-established fact that hierarchies among female macaques are virtually independent of weight, physical condition, and other indicators of fighting ability. The status tradition is primarily a *social* institution. Juvenile members of high-ranking lineages behave dominantly only when their relatives are nearby; their rank depends on the presence of supporters, rather than on some inborn predisposition.

Another indication of the importance of social as opposed to genetic factors comes from infant switching, practiced at the Wisconsin Primate Center to prevent inbreeding. We occasionally add fresh blood to a group by replacing a female's own infant with another one born at the center in the same period. If done within the first couple of days and with same-sexed infants, no problems ensue. At present, three such adopted monkeys of adult age live in the rhesus group. Each has the status that one would predict for legitimate offspring. One infant, named Orkid, was adopted by the alpha female, Orange, and is now the second-ranking female of the group.

Young females have to wage countless battles before settling at the predestined status level. It is no easy task to break the resistance of the heavier and stronger adult females of lower-ranking lineages. The youngsters consistently receive the necessary support from relatives, but there is also evidence that the female community as a whole supports the kinship system. Jeffrey Walters, in his fieldwork on baboons, found that adolescent females are backed not only by kin, but also by lineages ranking above their own. He noted that "animals were extremely reluctant to intervene against the existing or expected hierarchy." This is an interesting difference from humans and chimpanzees, who frequently do stand up for the underdog. In this way some of the "injustice" inherent in hierarchical systems is corrected, leading to a more flexible, more democratic structure. By comparison, the society of rhesus monkeys is strikingly undemocratic.

Help from "outside the house" sometimes enables young females who have lost their mother to claim the rank expected had she survived. Walters gives a few examples, and we have one of our own as well. Ropey is the only captive-born monkey without any relatives in the group. Before she was three years old, her mother died. Nevertheless, she has achieved the same high rank her mother had, just below the influential O-family. How did she do this? Not on the basis of physical strength; Ropey is only a little heavier than 5 kilograms, whereas some of the females she dominates weigh 9 kilograms or more. We suspect that her position is based on the backing she receives from the O-family and the B-family, which ranks just below her. Someone unaware of Ropey's history would certainly mistake her for the head female of the B-family, for she has strong ties with them. Ropey and Beatle, the true head female, are inseparable friends who spend more time together than any sister combination in the group.

Ropey's high position could perhaps be explained on the basis of genetic influence; she may, for instance, have inherited her mother's strong personality. Whereas, for reasons given before, a majority of primatologists believe that the social environment holds almost all the answers to the question of rank transfer among female macaques, breeding experiments are needed to look into the relevance of genetic factors. These are long-term projects, of which several are now under way. The results are anxiously awaited, because the "blue-blood hypothesis" is quite controversial—and not only within biology.

Aggression Levels

Aggression is a conspicuous aspect of the social life of rhesus monkeys. It is part of their hotheaded, belligerent temperament. The frequency and fierceness of attacks among these animals is amazing. Wild rhesus monkeys show scars and scratches, frayed ears, stumpy fingers, and other signs of intense fighting. Irwin Bernstein and his coworkers at the Yerkes Regional Primate Research Center in Atlanta reported for their rhesus group an average rate of eighteen aggressive acts performed per monkey per ten hours of observation. The group

Among rhesus monkeys physical violence is not, as among most other primates, exceptional. A dominant female bites her victim in the back while pressing her down on the feeding floor, which is covered with pellets of monkey chow. (Wisconsin Primate Center)

lives at a field station, in a large, open enclosure, with a surface area twenty times that of the monkey cage in Madison. The number of monkeys in our group is smaller, and our cage has a more vertical structure; still, the population density is much higher than in the Yerkes group. Observing our monkeys with similar methods, we found exactly the same aggression rate of eighteen acts per ten hours. When we limited the comparison to physical forms of aggression (slapping, grabbing, biting), we found no difference either.

The similarity in aggression level even includes free-ranging rhesus monkeys. Jane Teas and her colleagues studied a population of nearly seven hundred monkeys who roam the grounds of two ancient temples in Kathmandu, Nepal, feeding on offerings left by worshipers. The investigators analyzed their data by sex, reporting sixteen aggressive acts per ten hours carried out by the average female and thirty-eight by the average male. Only one of our males reaches this high a level, and the average for females in our group is very close to that determined by Teas. The consistency in the results of the three studies is almost too good to be true. Yet the methods for documenting monkey behavior are so well standardized that I trust the outcome, even though it completely refutes the popular notion about the effects of crowding. In stable, provisioned groups, spatial restriction seems to have no effect *at all* on aggression. The conclusion is that group life generates a fixed amount of friction among rhesus monkeys, whether they live in an open field, a large corral, or a cage.

Obviously, this holds true only within reasonable limits; if monkeys are packed in tightly, aggression is bound to get out of control. Yet this is not the condition in Madison. Approximately fifty monkeys live on a semicircular surface of 100 square meters, enlarged by a high rock structure. The greatest distance from one point to another is 15 meters, the maximum height 6 meters. In the same building three other monkey groups are housed, and on the second floor, windows give researchers a view from above. Our observation program is extensive. It involves compiling long-term records of global categories of behavior, such as grooming and coalitions. Most time-consuming of all, we take hundreds of so-called *focal observations* in which we concentrate on one individual at a time, reporting all his or her social interactions on a cassette tape recorder.

Watching animals is only part of the work. The observations have to be transcribed into computer files and compiled into tables and graphs before they can be interpreted. Behavioral study has its tedious side, as does any branch of the natural sciences. Studying more than one group, Lesleigh Luttrell and I recognize well over a hundred individuals. Chest numbers do not help much during observation; in fact, on moving or huddling monkeys they are unreadable. We tell the monkeys apart by differences in face, size, and color. I realize that to many people all monkeys look alike, yet the more familiar one becomes with them, the more distinctions one sees.

The reproductive cycles of rhesus monkeys, both in the natural habitat and in captivity, are synchronized in distinct seasons. From September through December all females come into heat, recognizable by the scarlet color of the skin on the bottom and legs. This is a very busy period for both sexes. The alpha male, Spickles, loses weight each mating season, dropping in four months from 13 to 9 kilograms. He is a magnificent old male, of perhaps twenty-five years, who normally moves slowly and on cold, wet days seems to suffer from arthritis. In the mating season, however, he tries to keep an eye on everything that is going on, and as if that is not enough, courts the females living in the adjacent enclosure as well. He watches them from underneath the door, sending his special courtship look, with protruded lips.

Aggression occurs in all seasons. Males become competitive in the mating season. Mothers defend infants in the birth season. Yearlings are forced to grow up when their new siblings are born, and they start challenging females of low-ranking families.* One can calculate what a rate of eighteen aggressive acts per monkey per ten hours means for a group of fifty: one and a half acts per minute. This figure needs to be qualified, however. In the first place, aggression usually occurs in bursts, involving many individuals at once. There are long stretches of time without quarreling in the group, then a few moments of

---

*It should be noted that the primate species treated in this book differ dramatically in their speed of development. *Infancy*, the period of total dependence on the mother for nourishment and transport, lasts approximately one year in macaques, compared to as much as five years in great apes such as chimpanzees and bonobos. The playful stage of *juvenescence* lasts until the age of three in macaques, as opposed to eight in apes. This is followed by *adolescence*, a period of sexual maturation and growing independence, which varies greatly in length per individual. If we accept full-grown body size as the criterion of *adulthood*, macaques reach this stage at approximately seven years, apes at about sixteen years. Reproductive ability is fully developed several years before that, however.

Two six-month-old monkeys. Because rhesus monkeys are seasonal breeders, their offspring always have playmates of the same age. (Wisconsin Primate Center)

massive activity. These complex outbursts of chasing and screaming do not make sense to the untrained eye, but they are highly structured by the existing hierarchy and the network of supportive relationships. Second, most aggression comprises mere threats. High-intensity attacks, involving fierce biting, occur in the group on an average of once every three hours.

The rhesus monkey has two forms of threat. One, with wide-open mouth and staring eyes, is commonly used by well-established dominants. The other, with ears flat and chin thrust forward, is of a less confident nature. Accompanied by noisy grunting, it is typically used by challenging adolescents. The response, especially to the first type of threat, is mostly flight. In addition, subordinates scream while baring their teeth in fear. This grimace can also be given silently. It may look to us like a friendly grin, but it is instead a very nervous one. The signal is typically given by subordinates as they are approached by a dominant.

Outcomes of fighting and competition are predictable, but not certain. An individual who wins most of the time against a particular opponent can therefore be called dominant, but there are exceptions. Reversals are usually brought about by influential third parties. For example, when a female becomes sexually attractive, a male friend may support her against females who normally dominate her. The grin face, in contrast, is *completely* consistent in its direction between individuals. If during a given period monkey A grins to B, B will never during the same period grin to A. In order for a signal to be immune to fluctuations in the social situation, it must be based on something very fundamental. My interpretation is that this facial expression is used to acknowledge the existing hierarchy. A particular individual may win one of ten confrontations with another, but both of them know very well what is the norm and what the exception. The monkey who mostly loses evaluates his or her

Rhesus monkeys typically threaten by staring with open mouth and ears moved forward. These two females, kept outdoors on a farm in Wisconsin, defend their sleeping cage (*background*) against an approaching human.

rank as inferior, communicating this by means of submissive grins whenever he or she meets the other.

We use the direction of this conspicuous signal as the criterion of formal rank. It is evident that dominance and power are usually in the same hands. Among rhesus monkeys there is not nearly as much room for social manipulation and influence from below as there is in the chimpanzee hierarchy. Hardly any primate, perhaps hardly any mammal, enforces status differences as rigidly as the rhesus. Dominants notice the slightest insubordination, correcting it by means of a threat or punishing

A juvenile rhesus monkey in the Wisconsin farm group (*photo at right*) grins at an intimidating adult male. It has been speculated that this facial expression evolved from lip retraction in reaction to noxious stimuli. The original reflex is shown (*above*) by a cactus-eating baboon (Gilgil, Kenya). In social situations the grin signals submission and fear; it is the most reliable indicator of low status among rhesus monkeys. In other species, such as humans and apes, this facial expression has evolved into the smile, a signal of appeasement and affiliation, although an element of nervousness remains.

it by means of an attack. They no doubt agree with Machiavelli that, if a choice is necessary, it is better to be feared than loved. For their strictness in these matters, I have even heard rhesus monkeys called the "chickens of the primate world." Nevertheless, studies focusing only on their peck order do not do justice to the species.

Evidence that rhesus monkeys are not all nastiness comes from an experiment by psychiatrists Jules Masserman, Stanley Wechkin, and William Terris. Several monkeys were trained to pull chains for food. After they had learned this response, another monkey was placed in an adjacent cage; pulling the chain now also caused the neighbor to receive an electric shock. Rather than pulling and obtaining the food reward, most monkeys stopped doing so in view of their mate's suffering. Some of them went so far as to starve themselves for five days. The investigators noted that this sacrifice was more likely in individuals who had themselves once been in the other monkey's unfortunate position.

This result may be contrasted with Stanley Milgram's famous experiments, in which humans delivered electric shocks to others. They were given the task of punishing fellow subjects for incorrect responses to a test. The victims were not actually wired to the electricity, or they might not have survived. They faked protest, though, by crying, pounding the walls, or begging to have the procedure stopped. As it turned out, many people are prepared to provide others with shocks of up to several hundred volts, labeled on the generator with DANGER: SEVERE SHOCK. The difference from the rhesus study is that the human subjects had been misled: they had been told that the purpose was to study the effects of punishment on the other person's memory, whereas the real purpose was to see how obedient they themselves were. The experimenter was constantly present, acting as an authority. The humans simply did what they were asked to do, assisting in the research. This soon became known as the Eichmann experiment, after the Nazi who killed hundreds of thousands of Jews but claimed that he had only been an instrument in the hands of others.

We tend to underestimate the extent to which rank and authority affect our behavior. Each year Elliot Aronson describes Milgram's experiment to his class of psychology students, asking them if they would obey. Only 1 percent indicate that they would, a figure about sixty times lower than was found in the studies by Milgram and others. Rather than believing that his students are better than the rest of humanity, Aronson concludes that words and deeds do not always agree.

The Exploratory Phase

Some modern textbooks have it that science starts with a battery of hypotheses, dispassionately tested for acceptance or rejection. I believe that science starts with fascination and wonder. Charles Darwin did not sail away on the *Beagle* to test a theory; he returned with the ingredients for one. The exploratory phase, as it is called, is indispensable to creative research. The first tasks of an ethologist beginning research on an unfamiliar species are to try to get into its skin; to think at its level; and, as stressed by the maestro of observation, Konrad Lorenz, to actually love the new species. In 1981 I turned "rhesus positive" and spent months immersing myself in this primate's hectic life-style.

What struck me most, after years of chimp watching, was the speed and directness of rhesian behavior. In the great apes there is a marked delay between impulse and action. Chimpanzees carefully take in the entire situation before making a move. They also seem to hide their intentions, which has earned them a reputation for unpredictability and deceitfulness. This is not true of rhesus monkeys; their emotions are very much on the surface. Rhesus seem a virtually transparent species.

Rhesus monkeys are definitely smarter than the average pet. Their intelligence is not of the obvious type, however, such as the apes' ability to piece sticks together to reach a banana. Rhesian intelligence is evident in the practical details of everyday social life. For example, a female named Beatle climbs a rock formation to join a huddle with her two sisters. Then she discovers the second-ranking male, Hulk, sitting directly behind them. She hesitates; Hulk's reactions are unpredictable. Beatle descends and starts feeding on the chow pellets strewn on the floor. Minutes later, Hulk comes down, foraging not far from Beatle. When she notices him, she at once glances from Hulk to her sisters and hurries off to join them. She has made a simple deduction: "If he is here, he can no longer be there." When watching closely enough, one can "see" monkeys think all the time.

Individual identities are crucial. There never seems to be any confusion about who is who. Rhesus monkeys keep track of all major events involving their kin and friends, as well as those involving their enemies. A few days after having given birth to her first infant, Ropey is huddling with Beatle. The baby, still very small, is completely hidden between the two females. Ommie, a close friend of both, approaches, looking a bit puzzled. She pulls Ropey's leg away to peek between her friends. Seeing the baby, Ommie releases the leg, lipsmacks,* and puts an arm around Ropey. Then all three of them lipsmack and huddle close together. Ommie apparently remembered that Ropey had an infant, and checked whether or not it was still there.

On the large feeding floor Orange makes two unexpected attacks. First she bites a young female named Tuff, and a little later she grabs Beatle. Usually the reasons for such outbursts remain unknown, but in this case there are some hints. I had been following Orange for half an hour and recorded the following events more than twelve minutes before the aggressive incidents: "Orange grooms Spickles. Tuff tries to sit against the old male's back, but he repeatedly shoves her away till she finally leaves. After this, Beatle comes over to groom Orange. She is not successful either, because each time Orange feels hands moving through her fur she turns around to threaten

*Lipsmacking is a series of rapid lip and tongue movements carried out by an individual with brief glances at the partner. Rhythmic smacking is most commonly heard during grooming, but it may also be performed at a distance, accompanied by raised eyebrows, as a visual signal of friendly intentions.

Beatle. Five attempts are rebuffed before Beatle gives up." This record shows that both Tuff and Beatle had intruded upon contact between the two top individuals, and Orange, instead of taking action during the important tête-à-tête, kept the interference in mind and punished later.

Individual recognition, memory, and simple logic underlie another ability: a monkey's comprehension of social relationships in which he is not himself involved. This allows an individual, say Harry, in his dealings with Bob and Mike, to reckon with the Bob-Mike relationship. For example, if Bob and Mike are allies, it is advisable for Harry to be friendly with Bob in the presence of Mike. Whereas if the two are enemies, Harry can try to play one off against the other. Because Harry has to take three relationships into account, we speak of *triadic awareness*. A few illustrations of this ability serve to show that reconciliation behavior, which I shall come to, does not stand on its own; it is part of a whole package of remarkable skills. The rhesus monkey's peace strategies cannot be understood without an awareness of the general social sophistication of the species.

The most basic social connection is the bond between mother and offspring. The first sign that other monkeys recognize this bond is seen in a vocalization given specifically to newborns. We call it baby-grunting, but since it sounds like noisy throat clearing, it is also known as cough-grunting, chortling, or gurgling. While uttering a series of baby-grunts, the third individual looks alternately at mother and infant. If the infant is carried by the mother, the two directions of communication are hardly distinguishable. If the infant is walking free, on the other hand, the baby-grunter looks back and forth between the two, with some grunts to the infant, some to the mother. The third monkey never addresses the wrong female, no matter how far the infant strays or how many other individuals are around. The call's meaning is not clear, but the intent is undoubtedly friendly. We anthropomorphically interpret it as a compliment: "What a fine baby you have!"

Proof that monkeys recognize ties among others comes from an ingenious experiment by Verena Dasser. A long-tailed macaque named Riche lived in a large captive group from which she was isolated for brief periods to test her responses to color slides of her primate friends. During these tests Riche looked at three pictures simultaneously, each picture portraying one of the many monkeys of her group. The slide in the middle always showed an adult female. This female's offspring was shown in one of the other two slides. It was not easy to predict who her offspring was, as the picture could appear on the left or the right. Riche had been trained previously, with different pictures, to pick out matching pairs. In the same manner, her task now was to select mother-offspring combinations. She hardly ever made a mistake, showing that she saw a clear connection between two of the three individuals on the screen.

Could family resemblance between the photographed monkeys have been the clue? No, because resemblance between relatives does not change with the age of photos, whereas tests with old pictures turned out to be unsuccessful. This makes sense only if Riche operated on the basis of individual recognition. A monkey's appearance does change over the years, making identification harder the longer ago a portrait is taken. Dasser concluded from Riche's success with recent pictures that, after seeing who was who, she applied her knowledge of the group's social network to select mother-offspring pairs.

Since monkeys classify one another on the basis of family ties, I was especially interested to see how a natural adoption in our rhesus group would work out. After his mother's death a three-month-old male, Kashew, was gradually integrated into the H-family of matriarch Heavy (obviously a large female). Heavy's attitude to Kashew developed over time from grooming him and tolerating his presence, to holding him and letting him suckle (although she probably did not have any milk). It was three months, however, before we first saw Heavy carry her adoptive son or defend him against others. It was much longer, almost a year, before the other group members began treating Kashew as a member of the H-family. The first occasion took place when Orange and Ropey together attacked Kashew. Immediately afterward the two dominant females, side by side, charged at Heavy and her adult daughter, who had both stayed clear of the scene, to threaten them as well.

This phenomenon, the *generalization* of aggression against one individual to include the entire family circle, is common among rhesus monkeys. For example, one female, threatening and chasing another female, repeatedly walks over to her opponent's daughter to threaten her too. At first it is not certain whether she is generalizing, because another explanation is that this daughter might have approached or done something else to draw the aggressor's attention. But then the aggressor suddenly walks up to a group of huddling, sleeping monkeys and jumps into their midst to grab one of the innocents. It turns out to be her opponent's sister.

During a focal observation on an old matriarch named Nose, I see her adult daughter attacked on the floor by Hulk. Nose sits far away, high above the scene, next to my observation window. She wisely does not move. Then Hulk starts looking around, scanning the groups of monkeys gathered on the rock structure. Finally he spots Nose, jumps up, and chases her.

After a fight between two females, one of them finds and threatens four of the five family members of her opponent. The fifth member is the juvenile son of her opponent's sister. He hangs upside down on the roof, playing with his peers. Individuals in this position are hard to identify. A few minutes after the incident, however, the female rushes to the roof to chase the one relative she has missed.

The same tactics have been observed among free-living vervet monkeys and savannah baboons. After an encounter between male baboons it is not unusual for one of them to seek out the favorite female friend of his rival and discharge his tensions on her. Barbara Smuts noticed something familiar about this form of revenge: "If we cannot get X, then we go after someone who means something to X." The study on vervets in Kenya is particularly convincing, as it involves a large number of carefully documented cases. Dorothy Cheney and Robert Seyfarth found that if members of different families fight, the same two families often clash again later in the day, but—and this is crucial—not necessarily the same individuals. Their relatives have become antagonistic as well. Apparently these monkeys closely watch ongoing fights and resent the entire family of a monkey who has confronted their own kin. According to Cheney and Seyfarth, the spreading of tension from two individuals to the rest of their respective matrilines demonstrates that vervet monkeys have a detailed knowledge not only of their own relationships but also of the relationships of others.

In human society generalization of aggression is quite common, both on a small scale, comparable to the foregoing monkey examples, and on a larger scale. It takes on extremely dangerous proportions when entire religious or ethnic groups are blamed for the acts of a few among them. Within days of Indira Gandhi's assassination in India in 1984 by two Sikh bodyguards, the nationwide death toll of Sikhs killed by Hindu mobs rose to over a thousand. The useful, in itself harmless, ability to understand connections among others may thus be used to stigmatize, ostracize, or even eliminate innocent people.

Implicit Reconciliations

The rhesus monkey's family feuding system affects not only aggression but also peacemaking. In the first place, the unity of the family has to be maintained at all costs. Second, one or two important family members may set the tone for the relations with another family. If they are at war with them, the rest of their kin join in; if they make peace with them, the rest relax too and resume normal relations. I spent much of my time following this kind of process. Do rhesus monkeys reconcile after fights? It depends. If kissing and embracing are the criteria, these monkeys do not measure up to humans and chimpanzees. Intense reunions between former opponents do occur, however, especially after serious tensions within a kinship unit or among close friends.

Orange's two daughters, Ommie and Orkid, have a fierce fight that soon involves the entire family. All the O's have their hair out. Orange herself sides with Orkid, the younger daughter. Not satisfied with biting each other, the sisters also discharge their tensions by threatening bystanders. The episode ends with an assault by Ommie and Orange on an old female, during which Ommie lipsmacks to Orange. I start my stopwatch when the fight is over.

Within one minute Ommie and Orange move around each other. Ommie presents her bottom to her mother but is ig-

An emotional reunion within the O-family after a serious fight between sisters Orkid (*left*) and Ommie (*right*). Orange, sitting between her two daughters, utters friendly grunts, while Ommie lipsmacks to Orkid. Orkid, in turn, lipsmacks to Orange's infant. Even though the females are oriented toward each other, they avoid direct eye contact. (Wisconsin Primate Center)

nored. Then, starting very carefully, she grooms Orange's back. During the second minute she is joined by Orkid, her main opponent, who grooms her mother from the other side. Soon afterward the ice breaks. The three females embrace and lipsmack in almost convulsive bouts. Normally, this lasts only a few seconds; this time, after each pause, one of the three females starts all over again, joined by the others. The O-family lipsmacks for two minutes. The old female who was its last victim gives a few lipsmacks from a distance. It takes twenty-one minutes before this female dares to join the huddle. The O-family stays together for no less than forty-three minutes.

The two middle-ranking families, the G's and the T's, often wage protracted, nonphysical battles with undecided outcome. On one occasion the G-matriarch, Gray, has been chasing a much bigger T-female, Tail, who seeks refuge near Orange. Tail recently has had an infant, which allows her to make contact with top-ranking females. Orange's presence clearly inhibits Gray's aggression. Gray sits down not far from Tail, grooming herself and glancing repeatedly at her opponent. The grooming activity seems to calm Gray. More than a minute passes before she looks up again; but now Tail has disappeared! Gray stands bipedally in order to get a good overview of the many huddled groups around her on the floor. She even moves on two legs from one group to another, systematically searching. Eventually she finds Tail sitting with her mother. Gray walks over and stretches out in front of her former adversary, with her back turned toward her. Both Tail and her mother accept this invitation and groom Gray.

Not all reconciliations occur so soon after the conflict. For example, Hulk chases and bites the third-ranking male, Mopey, normally his best buddy. Afterward they sit on bars attached to opposite walls, as far apart as the enclosure permits. They have their backs to each other. As far as I can judge, they manage not to look at each other for more than an hour. Then Hulk approaches, and the two males mount each other in turn. Younger males hurry over so as not to miss the reunion. They sit with Hulk and Mopey in what we call the male club. (Starting at a young age, our males huddle separately from the females, all together in a single cluster. Only Spickles never joins them.)

Orange's daughter Ommie has reached the age to challenge adults of the high-ranking B-family. She chases and grabs Boss, who successfully fights back until Orange intervenes. From Boss's subsequent behavior it is clear that she cares more about her relationship with Orange than with Ommie. Boss ignores the younger female and stays within a few meters of Orange, no matter where she moves, for the entire afternoon. Boss lip-smacks to Orange from a distance, gives loud baby-grunts when Orange's youngest offspring walks loose, threatens low-ranking monkeys while soliciting Orange's support by present-

An adult female (*right*) lifts her tail to present her bottom to the second-ranking male, Hulk, who has just chased her around the cage. The male ignores her overture, but later he allows her to groom him. A youngster follows the scene—and may learn from it. (Wisconsin Primate Center)

ing her bottom to her, and so on, for at least three hours. I go home without seeing a contact between the two females. Probably Boss has good reason for being so careful in her approach: fighting back against an O-member contradicts all the rules, and Orange is not known for her lenient personality.

During the birth season the presence of infants facilitates contact between female opponents. Although Boss did not succeed with her baby-grunt tactic in the above story, sometimes it does work. For instance, Orange threatens and chases Heavy. Heavy returns, giving baby-grunts to the alpha female and her infant, who is climbing the wire mesh. A little later Heavy's own infant comes close to Orange. Now it is Orange's turn to baby-grunt. This is the signal for Heavy to approach. Both females turn the wriggling youngster upside down, inspecting him closely while exchanging lipsmacks and baby-grunts. Their tensions are forgotten.

Almost accidental reconciliations are typical of rhesus monkeys; they often act as if nothing has happened. This impression results from their tendency to look in all directions except in the face of the former opponent. The human observer is apt to get quite confused because the rules for eye contact among these monkeys are so different from our own. Both humans and apes avoid eye contact during strained situations and seek it when ready to reconcile. Rhesus monkeys, in contrast, look each other straight in the eye during conflict; dominants intimidate subordinates by fixedly staring at them. Since prolonged eye contact is ominous in their communication, it is logical that they carefully avert their gaze during friendly approaches, including reconciliations.

The result is all kinds of "excuses" to approach the other party after a serious confrontation. For example, Spickles chases Hulk without catching him, because Hulk is much faster than the old male. When Hulk takes a sip from the water basin five minutes later, Spickles at once moves over to drink with him, their heads touching.

Boss has threatened and pushed her friend Tip out of a huddled group. After the incident she approaches Tip several times, but each time her opponent withdraws. Boss starts to hunt for flies, grabbing at the air with fast, snatching hand movements. This is a common technique. While so occupied, Boss moves closer and closer to Tip, without looking at her. She catches flies in front of Tip and behind her back. At one point she has to lean on Tip to get at a particularly high-flying insect. This contact is maintained and eventually leads to Tip's grooming Boss.

The most common "excuse" is the so-called contact pass. One individual purposely moves from point A to point B in the cage, "finding" his or her former opponent on the path. Tip chases Kopje, the lowest-ranking female, along the cage ceiling. Kopje is in the final days of a pregnancy. After the pursuit she sits still for six minutes, panting heavily from her strenuous escape. Tip moves to 1 meter above Kopje on the rock formation. Kopje glances over her shoulder at least twenty times a minute to check Tip's whereabouts. Then Tip descends, passing Kopje so closely that their hairs brush. This contact immediately relaxes Kopje. She descends too and feeds on the floor not far from Tip.

Contact passes seem to carry a soothing message, not because of what happens but because of what does not happen. During the pass the dominant can easily grab the subordinate. Instead, the dominant peacefully continues on his way. The same message of "Look, I'm not going to hurt you!" is conveyed by sitting down in full view, very close to the subordinate for a brief moment before moving off again. Such interactions are perhaps more accurately described as tension breakers than as reconciliation. Rhesus monkeys, in my opinion, are not very good at reconciliation, but they do have many subtle ways of letting one another know when a conflict is over.

Both levels of making up are observable in humans. I call them *implicit* and *explicit reconciliations*. The first type, in which no reference is made to the previous conflict, is rhesuslike. You meet your colleague, with whom you had a spat yesterday, and act as if nothing had happened. You get him some coffee, comment on the weather, or start a work-related discussion. The colleague does not mention the incident either, yet by responding normally—not too coolly and not too enthusiastically—he shows himself to be without rancor, or at least willing to act that way for the moment.

I once witnessed a dramatic case between two women at a scientific meeting. One of them had led a workshop during which, after a more-than-lively group discussion, she took a younger woman aside to reprimand her for getting too excited. The second scientist felt extremely humiliated. She was so upset that she was sick to her stomach, and she looked pale and withdrawn for the rest of the day. She seemed in better condition when I met her the next evening at the square of the little German town. We walked together. I apparently have developed an eye for this sort of situation, because I was the first to notice her opponent approaching in the distance, busily talking with other people. The two met in the middle of the street—the tension only visible, I suppose, to those who knew what had happened between them. The older woman came closer, bending to touch the younger's colorful belt. It was splendid, she exclaimed. Their eyes had hardly met before this brief grooming contact. Afterward they chatted about restaurants and other trivialities, at first uncomfortably, later in more relaxed fashion. Although their earlier collision was not part of the discussion, it must have been at the back of their mind the whole time.

An explicit reconciliation is one in which the parties *do* mention the previous conflict. They apologize or try to eliminate any misunderstanding. The exchange may look like a renewed conflict, for the old disagreement is never completely dead.* Typically, in an egalitarian relationship, a compromise is reached with both parties sharing responsibility. In a relationship with a strong status component, however, the subordinate usually accepts most of the blame. If not, the conflict is likely to escalate as the dominant sees his or her authority in danger.

---

*Residues of antagonism can be expressed nonverbally in inhibited attack patterns. People—children particularly, but also intimate adults—may give each other a shove, a punch, or a light kick in the leg during rapprochement. The gesture seems to be made in jest, but the message is "This is what I would like to do to you!" In monkeys and apes similar mock punishments can be observed.

It is evident that the highest degree of explicitness is reached in our own species, for we alone have language to discuss the matters that divide us. Yet if two particular chimpanzees, who normally never kiss and embrace, do exactly that a short time after a major fight, it is hard not to see it as an explicit act of making up. They do not need to allude to what has happened between them; their behavior is so exceptional that it acts as an unequivocal reference to the past. In this sense it is different from the majority of contacts in the rhesus group, in which former adversaries behave casually or develop excuses for approaching the other. It must be added that people as well, whatever their abilities, seem to choose this implicit procedure most of the time. It is less embarrassing—and as long as it works, it suffices for many of our relationships.

Hard Evidence

Let a trained bear wrestle with a man and you have an anecdote. Let it happen hundreds of times, using different bears and different men, and you are ready for a decisive comparison of fighting abilities. Science draws a sharp line between anecdotal and hard evidence. An anecdote is a unique observation. It is a striking glimpse of some suspected phenomenon without, however, any conclusion that the occurrence was more than chance. Hard evidence is the result of repeated observations under a variety of circumstances.

In explaining how evidence is collected, I am entering what outsiders often consider the boring part of science: the statistics, controlled variables, alternative hypotheses, and so on. The monkeys become impersonal objects of investigation—stripped, in a sense, of their flesh and blood. But the quest for abstract truth also has an exciting side. It forces us to make our assumptions explicit and to take a critical look at initial interpretations. It is a challenge. If monkey watching is like staring at the moon, monkey research is like going there.

The foregoing stories of generalization and reconciliation in rhesus monkeys fall somewhere between anecdotes and hard data. The observations are not unique, because they describe events that occur continually. On the other hand, they do not prove very much. For instance, contacts following aggression may look accidental because they *are* accidental. So the next stage is more systematic observations—including control data, which are crucial in science.

For my control data I needed to compare contacts among former adversaries with normal contacts. "Normal" activity is commonly measured by watching animals according to a fixed time schedule. For my purposes, however, it was better to carefully match every observation made after an aggressive encounter with a control observation. Assume Ropey and Heavy have a fight at 2:10 P.M. Their behavior would be recorded twice: first, for ten minutes immediately following the incident; then, on the next day, again for ten minutes starting again at 2:10 P.M., but this time with no aggression beforehand. The advantage of this procedure is that the control data concern the same individuals, at the same time of day, during the same season of the year. It is more likely, therefore, that differences between the two data sets result from the specific factor in which we are interested: the previous occurrence of aggression.

This project kept a technician, Deborah Yoshihara, and me occupied for many months. We made the two types of observation on nearly six hundred pairs of opponents. Their behavior could be affected in three ways by the earlier conflict:

Dispersal. The traditional idea of aggression is that it causes animals to avoid one another, leading to dispersal. If this were true, we would find less contact after aggressive conflicts than during the control observations.

No effect. The so-called null hypothesis says that my ideas of reconciliation are all imaginary. Under this hypothesis we expect no differences between the observations.

Reconciliation. The third possibility is that rhesus monkeys seek reconciliation, or at least some form of tension reduction. If this were true, we would expect individuals to have more contact immediately after their conflict than during the control periods.

The first two hypotheses were proved wrong. Aggression was often followed by contact: 21 percent of the pairs engaged in friendly contact after their conflict, whereas only 12 percent did so in control periods. Opponents failing to make physical contact sat near one another more often than usual. It was quite remarkable to find so many former opponents close together, as conflicts consisting in mere threats had been excluded from the study. All encounters involved some amount of chasing, which is, in the short run, not exactly a behavior promoting proximity.

Are these results strong enough for us to conclude that reconciliation occurs? There is one hidden danger. In theory it is possible that aggression causes a wave of friendly contacts, involving many group members, regardless of whether or not they were adversaries. Because of its random nature, we would not want to call such a general change in activity a reconciliation. This theory was refuted, however. Our data demonstrated that contacts occurred *specifically* between the antagonists. Our conclusion is that rhesus monkeys are attracted to individuals with whom they have had an aggressive encounter. It is not just a matter of seeking calming contact with conspecifics; the former enemy is the preferred partner.

To my surprise, we found the same sex difference as in chimpanzees: male-male and male-female fights were more often reconciled than fights among females. It was not so much that the high scores of the males were unexpected—what bothered me was the low scores of the females. Remember the explanation of the sex difference in chimpanzees. Male chimpanzees cannot afford to hold grudges; in a highly competitive system of flexible coalitions, they must stay in touch with both rivals and friends. Female chimpanzees lead a more solitary life, committing themselves to their offspring and a few good friends; they can be more selective in their peacemaking efforts. The first piece of the argument is, with some modification, applicable to male rhesus monkeys as well. But I do not see how the second can hold for rhesus females, who organize themselves in such large, cohesive groups.

An alternative explanation of the sex difference is that it has more to do with rank than with gender. Maybe the mending of disturbed relationships occurs especially at the top of the hierarchy, where the risk when tensions escalate is greater. The fact that males usually occupy the high ranks would make it seem as if reconciliation were sex linked. I planned an experiment to tease apart the influences of rank and gender. For this I needed a large number of new monkeys, as it is our policy not to experiment with the breeding group.

The new individuals for my study knew the fundamentals of social life. The standing rearing practice at the laboratory of the Wisconsin Primate Center is to let young monkeys spend their first nine months in a group with their mother and other female-infant pairs, after which they are housed in so-called peer groups with monkeys their own age. I borrowed part of my experimental design from a colleague, David Goldfoot. Both he and the director of our center, Robert Goy, have been working for many years on the hormonal and social origins of gender differences in monkey behavior. One of these programs compares heterosexual and isosexual groups—that is, social groups composed of both sexes and groups that comprise only males or only females. The latter situation reveals how females behave in a hierarchy not dominated by males.

I created six isosexual groups of four monkeys each: three male groups and three female groups. Group members were strangers to one another. Since adults tend to fight too much when first put together, I used monkeys less than three years of age. Group formation was similar for both sexes. Within minutes, two of the monkeys formed a coalition against the rest. The second-ranking individual was usually the most aggressive. He or she spent much of the first days protecting the coalition by jealously interrupting all friendly contacts and play between the alpha monkey and the others. After stabilization of this situation, by the end of the first week, I began to experiment.

Working with one group at a time, I threw a quarter of an apple into the pen and recorded subsequent behavior for half an hour. In control tests I did everything the same way—entering the room, opening and closing the cage door, sitting down—except that the monkeys did not receive any extra food. My idea was to create a brief moment of tension and competition in order to see whether it was followed by an increase in positive behavior, such as grooming, play, or huddling. I expected such restorative behavior to be more common in male than in female groups, *unless* rank differences had been responsible for the sex difference found earlier in the large mixed group.

The initial response to the apple piece was the same for all groups: aggressive competition. Over 95 percent of the conflicts consisted in mere threats and chases. Remarkably, the apple piece was forcibly claimed by every dominant from every subordinate, except by alphas from second-ranking monkeys. As a result of this inhibition, success in food competition was equal for the top duos. The position of the alpha monkey appeared to depend on coalition with the second-ranking monkey, which meant that the two had to be careful to stay on good terms. Too much selfishness on the alpha's part might frustrate his or her partner and endanger the supportive relationship. This seemed a simplified version of the problems encountered by chimpanzee Nikkie and his cunning partner, Yeroen.

One alpha male, Dick, unsuccessfully tried to obtain the food by a trick. When Victor, his ally, had obtained the apple piece, Dick kept following him around, threatening but not attacking him. After four minutes Dick seemed to have given up. During the sixth minute he lipsmacked to Victor, who had started eating. Dick, tail in the air, presented his bottom. Victor responded as usual to this friendly invitation: he mounted Dick. Once Victor was on him, however, Dick abruptly spun around and grabbed for the apple piece. In the brief struggle Victor was able to keep the food, and Dick could only lick his fingers.

The outcome of the many apple tests that I conducted confirmed my expectation. After the initial aggression and the consumption of the apple piece, the males spent a lot of time together. They were actively clearing the atmosphere, showing greater cohesiveness and more grooming than during control tests. This was not the case in the female groups; females had rather less contact than normally. These results gave no support to the idea that rank, not gender, might be the crucial factor. For the present, I conclude therefore that there is a genuine sex difference in the psychology of peacemaking in rhesus monkeys.

Class Structure

In nature, group membership of macaque males is of the revolving-door type. Males come and join a group, stay for a couple of years, then migrate to another group, or live alone for a while. Even though his entry into a group is often resisted, the newcomer needs to build good connections with its core members, both male and female. In order not to end up at the bottom of the totem pole, he must do two things simultaneously: make friends and assert himself. These objectives are hard to combine without alternately hitting and hand shaking. Although this problem is different from that of the chimpanzee male (who stays in his home group), it involves the same sort of opportunism. Rivals need to be approached, victories need to be followed by appeasement. The conciliatory attitude evident in our study reflects this heritage of the migrating macaque male.

We still need to determine how rhesus females manage to live in such a highly organized society without putting much energy into reconciliation. The answer, again, is that females do not lack the skills for making up, but they use them more selectively than males. One indication is the high number of reconciliations among mothers, daughters, and sisters in the large group. But I did not fully realize the extent to which females channel their peace efforts until we began the so-called drinking tests.

Classic dominance tests start with depriving animals of food or water for twenty-four hours or more. After this they are provided with a single source, such as a drinking nipple, that can be monopolized by one individual. Obviously this arrangement creates an extremely tense, intolerant atmosphere. One by one the animals come to drink. The only thing the observer has to do is record the order in which they arrive. This convenient procedure has been criticized because it presents us with a one-dimensional view of social life. The neat hierarchical pattern that we see is our own creation, forced upon the animals by the situation. In the wild, food and water are distributed over space and time. Except during rare droughts, monkeys do not have to wait twenty-four hours to quench their thirst, and they drink side by side from puddles or streams.

I decided to mimic the natural situation in our large rhesus group. Research on reconciliations has its limitations; it can only provide information on individuals who regularly fight. Yet from the standpoint of peace strategy, the most interesting relationships are those in which aggression rarely occurs. In order to document those relationships I designed a new dominance test, which offered a choice between competition and social tolerance. I prefer this type of test because it does not put the group in as stressful and unsettling a situation as the classic test.

The water supply was turned off for only three hours, after which water was provided in a basin large enough for simultaneous drinking by four adults or eight juveniles. No orderly pattern of coming and going resulted. The monkeys arrived in ever-changing combinations, some individuals pushing aside others, some happily drinking together. Lesleigh Luttrell and I videotaped nearly fifty tests, with several thousand encounters between adults around the water hole. Four types of interaction resulted: two monkeys drinking together (26 percent); one monkey drinking, with another sitting nearby (15 percent); one monkey avoiding another (51 percent); and one monkey excluding another by means of aggression (8 percent).

The group's formal hierarchy can be divided in two; I designate the halves the upper class and the lower class. These terms do not imply that one category of monkeys is intrinsically better or superior, only that there is a difference in privileges. The origin of this difference is buried somewhere in the group's history. In our experiment all individuals of the upper class took precedence over those of the lower class. Within each class, however, the drinking order was virtually independent of rank. It was not at all unusual for the alpha male, Spickles, to arrive after half a dozen other upper-class monkeys. He had no trouble claiming the basin whenever he wanted; he simply seemed not to be in a hurry to exclude others. Sharing of the water basin was common among members of the same class, whether or not they were kin, but rare among members of different classes. Middle-ranking females seemed particularly unable to endure one another; females at the bottom of the upper class were very intolerant of females at the top of the lower class. Thus, the drinking order was flexible and easygoing within each class, and the classes seemed to be kept apart by competition between females at the class margins.

While a number of lower-class monkeys share the drinking basin, two watch for approaching dominants. (Wisconsin Primate Center)

These classes should not be regarded as separate subgroups. During associations and grooming the class division is not obvious at all; many female bonds cross the class boundaries. Perhaps the social structure resembles that of free-ranging Japanese macaques. Japanese primatologists distinguish a number of concentric rings around the heart of a group. In their terminology our upper class might be called the central part, and the lower class the peripheral part, of society. We at Wisconsin are the first to discover such a stratified rank order in a *captive* group of primates. Is our group unique? I find that hard to believe. It is more likely that a division in the distribution of social tolerance has escaped attention because traditional drinking tests do not leave room for anything other than competitiveness.

After becoming aware of this dividing line among our females, I returned to the data on reconciliations. To this point the comparison between the sexes had dealt with the aggregate of male and female conflicts. When I separated the female data by class structure, the sex difference almost disappeared. Conflicts within each female class, even among unrelated females, were reconciled as often as male conflicts. The likelihood of a friendly reunion was low, very low, only after attacks by upper-class females on lower-class females.

When we put the pieces of the reconciliation puzzle together, it is evident that there remains a gender difference. In new situations, such as the temporary groups created for the apple tests, males immediately begin to work on their relationships, smoothing things over after tensions. Females are probably more interested in long-term commitments, which take more time to develop—perhaps years. In a well-established social network such as the large breeding group, females concentrate on certain spheres of interest; they make up principally with their relatives and members of their own social class. So both sexes seem to do what serves them best in the natural situation, in which males wander from group to group and females stay in stable societies for their entire life.

The only part of this scheme that needs clarification is whether social classes really do represent "spheres of interest."

The intolerant behavior of females at the class boundaries reinforces this idea, but the real evidence has to come from observations of coalitions in which females support one another. Is there more solidarity within classes than between classes? Does the upper class operate as one solid power bloc against the lower class? According to our research, this is indeed the case. The selectiveness of female reconciliations is strategically motivated: rhesus females make up after fights primarily with individuals whom they need for cooperation in a competitive world.

When I first reported the remarkable social stratification of our monkey group at an international congress, one colleague publicly warned me about the term "social class." He questioned neither my results nor my conclusions, but was concerned that the terminology could be misused. Conservatively minded people might quote the observations to justify existing class differences in human societies—something along the line of "if monkeys form social classes, they must be natural." Marxists would be terribly upset, once again branding biology a reactionary science.

To prevent this I could, of course, speak neutrally of the upper and lower parts of the hierarchy, or simply of upstairs and downstairs monkeys. This terminology, however, would obscure the fact that one subgroup shares privileges in a relatively tolerant way, while excluding the rest of the group from these same privileges. Also, as indicated by our reconciliation data, there is more forgiveness among monkeys of the same category. Only the phrase "social class" captures these aspects of the situation. Not being allowed to use the most appropriate term is almost like having to describe the movement of birds as "going through the air" because the word "flying" has been claimed by some airline company. The flight patterns of birds and airplanes are not identical, just as class structures of monkeys and humans are not identical; but this is not sufficient reason to invent a different language for each particular case.

Rather than avoid so-called dangerous words, thereby leaving us with a depleted and meaningless vocabulary, biologists need to expose the error of simplistic political use of their findings. I would not be writing this book if I did not believe that the study of animal behavior sheds light on the roots of our own societies. It helps us place the human condition in perspective. Yet none of the many lessons to be learned provide us with *norms* for our own behavior. Humans have a lot of flexibility in the way they structure their societies, depending on the education they give their children and the laws and institutions they create. What is important is not whether our social institutions are "natural" (whatever that is), but whether they work well and benefit most people. Only by careful evaluation will we know what is best for us.

All this is said to make clear that I do not feel I am endorsing class structure in our own species by applying the term to rhesus monkeys. I do recognize similarities, though, such as the taboo on food sharing across classes in the traditional Indian caste system. The rationalization for this taboo was "purity"; that is, members of the higher caste risked being polluted by accepting water from the vessels of lower-caste members, by smoking with them, or even by standing too close to them. Also, violence against members of different castes was judged differently; penalties for murder ranged from twelve years to nothing, depending on whether you killed a Brahmin or an Untouchable. The virtual absence of reconciliation with lower-class victims in our rhesus group is a strong parallel.

Obviously, there also exist major differences, such as the religious and ideological frameworks that humans build around their class systems, the division of labor and the accumulation of wealth, and the possibility for humans to move up a class via marriage. This may be forbidden by law (as it was until recently in the South African apartheid system), but in most stratified societies interclass marriage is discouraged but not prohibited.

Climbing the Ladder

Of all monkeys born in a wild bonnet-macaque population in Sri Lanka, nine out of ten die before reaching adulthood. In wild rhesus monkeys the situation is hardly any better. Our much healthier caged monkeys may not live in a jungle paradise, but they need not be pitied. Freedom does not necessarily mean happiness. The astonishingly high death rates of monkeys in nature are due to starvation, disease, and predation. Survival chances are worst for offspring of low-ranking females. Being dominated and chased away from food sources causes a lot of misery, hardship, and stress for wild monkeys at the bottom of the hierarchy.

These results have recently been confirmed by observations on arboreal long-tailed macaques. Having visited the forest site in Sumatra where this study was conducted, I know how terribly difficult it is to watch these monkeys. The rain forest is dark, the canopy dense and high, and the brownish-green monkeys blend perfectly with their environment. Two Dutch biologists, Maria van Noordwijk and Carel van Schaik, spent years among the Sumatran macaques—and also among the orangutans, tigers, and millions of leeches there. They found that high-ranking monkeys are usually at the forefront when entering trees that are loaded with ripe fruits. These monkeys obtain high-quality food with less effort, allowing them more time to rest and groom. Subordinate females are forced to travel and forage away from their group's main party, which probably increases the risk of predation. These females often disappear for unknown reasons, and their infants have a low survival rate.

Being at the top of the social ladder is not merely a pleasant, comfortable position for a wild female monkey: it determines her life span and reproduction. The fact that dominant females succeed in rearing more offspring means that their genes spread through the population. Traits—such as social skill and ambition—that may have helped these females to secure a top position for their lineage are inherited by large numbers of monkeys. And captive rhesus females demonstrate this inheritance. They attach great importance to status, much more than one might expect in view of the abundance of food and the absence of predators. The consequences of millions of years of evolution are not erased in a couple of generations.

The hierarchy among rhesus monkeys is so strict that dominants may even claim the contents of a subordinate's cheek pouches. Here a juvenile submits to inspection. (Wisconsin farm group)

For males the situation is slightly different, because the advantages of high rank continue to be significant in captivity. It is generally believed that male dominance pays off in terms of sexual privileges. How this affects a male's reproductive suc-

A female (*right*) sleeps on one of the bars near the cage ceiling, not the most comfortable of places. Her upside-down daughter does not seem to mind. During periods of tension low-ranking individuals try to stay out of problems by isolating themselves. This family ranks at the bottom of the hierarchy. (Wisconsin Primate Center)

cess is difficult to ascertain. It is relatively easy to measure a male's sexual access to females. His actual reproduction, however, can only be revealed by means of paternity tests based on blood types and other genetic data. This method is now increasingly used, both in the lab and in the field. One of the first thorough studies involved our own group.

For nearly a decade the sexual behavior in the group was observed by researchers at the Wisconsin Primate Center. At the same time, Marty Curie-Cohen and his colleagues in the Genetics Department collected blood samples from all infants born into the rhesus group, as well as from their mothers and possible fathers. The number of adult males and their rank order varied over the years. In each mating season the most dominant male was observed to copulate much more frequently than any other male. However, this alpha male was not always the one who sired the most offspring. Upcoming young males of second or third rank often fathered more infants. Perhaps their sexual activity was more discreet. Or their sperm may have been more fertile than that of the older males. Whatever the reason, it appears that reproduction takes place predominantly via males who are either high ranking or show a potential for becoming so in the near future.

The present situation in the group is no exception. Spickles does all the "public" mating. He copulates in full view of everybody, uttering a series of loud barks at the climax. The number two male, Hulk, never mates when Spickles can see him and certainly does not call attention to himself when he does mate. His surreptitiousness is amusing to watch, especially in light of the political complications created by Orange's sexual attraction to him.

The group's top triangle of Spickles-Orange-Hulk can be summarized as follows. Mr. Spickles is a social institution maintained by the female population, headed by Orange. It is uncertain, in view of his age, whether Spickles can hold his own against younger males without female backing, so his ties with Orange seem crucial for the stability of his position. The two of them spend no less than 9 percent of their time grooming each other, an amazing level when we consider that the average

Orange grooms Spickles. The two alpha individuals spend an enormous amount of time together, ruling the group as a team. (Wisconsin Primate Center)

grooming time for other male-female pairs is less than 0.5 percent. The two alpha individuals also support each other in conflict situations: they rule the group as a team.

Yet their solidarity is not perfect. Orange supports Spickles against his principal rival, Hulk, as long as the old male only threatens or chases. On the rare occasions when Hulk has actually been attacked, Orange has defended him. Thus, she helps the number one male to maintain the status quo without allowing him to progress to the point of physical harm to the number two male. Understandably, both males seek good relationships with her, and although Spickles definitely is more successful at establishing contact, Orange does allow Hulk to spend a good deal of time with her and her royal family.

Since it is not in the nature of rhesus males to form lasting coalitions, Orange does not need to worry about a ruling unit that excludes her. As a consequence, she not only absolutely dominates the entire female population, she is also part of the central triangle and has the most leverage. Even in rhesus society the rank order does not completely reflect the power relationships: Orange is formally lower in status than Spickles (she gives fear grins to him and avoids his charges), but at the same time she may well be able to make or break him. I qualify that statement a bit because it is always hard to assess what is going on behind the scenes in a stable social situation. The exact nature of such influences will only be revealed when the position of one of the three top monkeys is threatened, from inside or outside the triad. To date this has not happened.

The only major instabilities in the troop hierarchy over the past six years have been caused by one middle-ranking female, Tip. At the end of 1981 Tip still regularly bared her teeth to her mother and her adult sister. In February of the next year things began to change. On several occasions, when chased by her sister, Tip appealed for help to high-ranking females and to Hulk. She would present her bottom to them while threatening and screaming at her opponent. Sometimes this led to a real fight, with Tip as the attacker. Twice her sister had to be temporarily removed from the group for treatment of injuries.

The family just above the T-family in rank became involved,

and relations between its matriarch, Gray, and Tip began to show serious strains. Gray frequently threatened Tip, especially when Tip approached one of her usual supporters. These problems were resolved by joint aggression against the sister of Tip, a discharge that had a strong unifying effect on Tip and Gray. They often broke off their involvement in the attack to groom each other while the poor sister was still being pursued by the supporters Tip had mobilized. Reconciliation with and support from Gray ranked higher on Tip's priority list than the well-being of her relative. During these months the two T-sisters never groomed or even sat close, and neither grinned submissively to the other.

Formal dominance was reestablished in July 1982, when Tip's sister bared her teeth for the first time. Although noisy conflicts remained a daily occurrence, physical fighting between the sisters disappeared. Tip's ambivalent relationship with Gray continued, however: at one moment they were on the verge of fighting; at the next, they were seeking a scapegoat. Tip now left her sister alone, concentrating instead on her mother. This led to fierce altercations in which Tip successfully recruited support from all the G-females. By the end of 1982 Tip's mother too formally submitted. My interpretation of this whole process is that Tip had smartly exploited Gray's tendency to "generalize." Gray's tensions with Tip were easily turned into aggression against the rest of the T-family, which was exactly what Tip needed.

Tensions within the T-family persisted for a few more months, but were put to rest in the spring of the next year. Arriving one morning, I noticed from a distance two females baby-grunting and touching each other's infant. Then they groomed each other. Nothing unusual for the average zoo visitor, but highly significant in my eyes. The females were Tip and her adult sister. Almost two years of ill feeling had come to

Mr. Spickles yawns, a sign of tension, as Hulk stands right above him. Hulk may be in better physical shape, but he lacks the experience and female support that keep Spickles firmly in power. (Wisconsin Primate Center)

Tip (*right*) shows the threat face typical of monkeys challenging the existing order. With ears flat and chin pointing upward, she grunts at her adversary. Over the years Tip rose in rank above a number of females, including her own mother, who screams in protest (*above*) after an attack by her ambitious daughter. (Wisconsin Primate Center)

an end. From then on, contact within the T-family gradually increased. By 1984 the T's formed one of the most cohesive matrilines of the group. With Tip firmly on top, peace had returned.

This is not the end of Tip's story. As was to be expected, Tip subsequently took on the dominant G-females, waging daily battles with them. Gray's oldest daughter had, and still has, a good deal of trouble with Tip and her powerful supporters. (Old Gray herself died in 1983 of natural causes.) In this struggle as well, Tip and her adversaries never bare their teeth to each other and never groom. It is interesting to compare this with chimpanzee behavior. The common point is that dominance struggles involve a disruption of the formal relationship, that is, an absolute cessation of status communication as long as the process is undecided. The great difference is that while rival

rhesus monkeys stop all friendly exchanges, reconciliations among the male chimpanzees in Arnhem were frequent during such tense periods, resulting in an actual grooming *increase.* (I do not refer to the recent fatal incident, but to the dominance struggles of earlier years.) This seems again a matter of conflict resolution and compensating measures having reached a higher stage in these apes than in rhesus monkeys.

Tip's ascent is remarkably slow, and it is unclear how and when it will end. On first thought, it may not seem to make much difference whether she will rank seventeenth or thir-

After Tip established dominance over her kin, peace returned within the T-family. Here Tip grooms her aging mother. (Wisconsin Primate Center)

teenth. But in light of the class structure, if Tip succeeds in outranking the G-females she will be in a position just below the bottom of the upper class, from which she can perhaps join that class. This would represent quite a major step in the peck order: both Tip and her offspring would enjoy the relative solidarity and tolerance of the group's elite. Of all lower-ranking females she has developed by far the best grooming connections with upper-class females. Somehow she seems to be making all the right moves. Several high-ranking males and females are aiding her cause, and even her sister has recently turned into a supporter.

Does Tip know what she is doing? Does she know where she is going? Although it is impossible to be sure, I am inclined to

Tip (*right*) grins submissively and withdraws in the face of open-mouthed threats from Orange and Ommie (*left*), whom she has tried to join. Tip's increasing association with top-ranking monkeys still meets with resistance. (Wisconsin Primate Center)

believe that she does. It is difficult to explain the flexible strategies of nonhuman primates without assuming that they are aware of the consequences of their behavior. This perspective may seem reasonable, yet the traditional view of animals is quite different. More than fifty years ago Solly Zuckerman noted that "subhuman primates have no real apprehension of the social situations of which they themselves form a part." American behaviorism in particular has contributed to a mechanistic picture of animals in science. Animals were viewed as furry billiard balls, blindly rolling around, their course determined by the laws of physics, or, in the terminology of behaviorism, by stimulus-response contingencies.

Stuart Altmann has applied this view to the dominance relationships of primates. Playing the devil's advocate, he has compared such relationships to the Cheshire cat's grin. They are an abstraction, he says, existing only in the mind of the investigator. "Are dominance relationships important? They surely are, but to the investigators, not to their subjects."

Let me summarize a few phenomena that may shed light on this problem:

• The fear grin is a signal no rhesus monkey ever gives to a subordinate in spite of the occasional loss of a confrontation. Rather than reflecting the outcome of transient conflict, this signal must depend on an evaluation of the long-term relationship. In other words, rank is not an abstraction; in their communication monkeys refer to the status differences among themselves.

• Females ranking near the boundary of their social class are remarkably intolerant toward females on the other side of that boundary. This suggests an awareness of the stratification of the group.

• Rhesus monkeys generalize from one to all members of a kinship unit. There are other indications of triadic awareness, that is, of an understanding of the relationships among others.

• Rhesus monkeys remember with which individuals they have fought. Whether or not they reconcile with them depends on affiliative ties and class membership.

Strictly speaking, the above list does not prove that monkeys know what their behavior is all about. The point is that their actions become more understandable if we *assume* that they have insight into the social network in which they live. Regarding them as beings with a rich social knowledge and a will of their own permits us to interpret data that otherwise would not make any sense. So I am talking here about a theoretical framework rather than a proven position. This framework, known as *cognitive ethology*, is more stimulating and promising than the classic view of animals as robot actors in a play that they hardly understand. Instead of arrogantly thinking that we human investigators fathom the meaning of nonhuman primate behavior better than they do, the impression never leaves me that it is the other way around. It takes me thousands of hours of waiting and watching to reach a depth of insight into their social life that, in my estimation, is shallow compared to the insight of the monkeys themselves.

On the other hand, there is no indication that monkeys comprehend the total outline of their social organization. They may know their social hierarchy intimately, but this is no guarantee that they have any conception of what a hierarchy is. "The integral picture does not exist in his mind; he is in it, and cannot see the whole from the outside," wrote Bronislaw Malinowski in 1922. He referred to the inhabitants of the Trobriand Islands, but I read his statement as applicable to monkeys, not humans. "They know their own motives, know the purpose of individual actions and the rules which apply to them, but how, out of these, the whole collective institution shapes, this is beyond their mental range." Undoubtedly the anthropologist was underestimating the Trobrianders' mental range. Humans can and do acquire an overview of their society, and there is no reason to believe that this capacity varies from one people to another. This is not to say that humans consciously take this overview into account in everything they do. On the contrary. Most of the time, we, like other primates, act on the basis of intuitive knowledge of our immediate social environment.

CHAPTER FOUR

Stump-Tailed Monkeys

The red face color, limited to the areas around
eyes and nose, looks peculiarly pockmarked and
darkly blotched, almost diseased. A laryngeal
sack with a wattle, and a fat, sparsely furred
belly are further "beauties", which make the old
male one of the ugliest of all primates. Even the
striking demoniacal character of the drill and
mandrill is lacking, while also the soul of the
bear-macaque does not show anything of
temperament and energy. Instead, they are quite
phlegmatic.

—Alfred Brehm

If I had to convince a skeptic of the existence of simian
reconciliations, I would take him to neither the chimpanzees
nor the rhesus monkeys. Chimpanzees, with their long memory span, take their time in everything they do. Untrained observers have trouble focusing on two former antagonists for
more than a couple of minutes and get distracted by unrelated
events. With rhesus monkeys, on the other hand, reunions
after fights are often too subtle for humans to grasp their meaning. We do have evidence for a connection with the previous
aggression, and I am persuaded that rhesus reunions are
loaded with meaning for the monkeys themselves; but unfortunately, as I said, my guest is a skeptic. So I would take him to
the stumptail group.

Our Beauties

Of the primates treated in this book, stump-tailed monkeys*
make peace in the most predictable and conspicuous manner.
I can guarantee the occurrence of a handful, perhaps even a
dozen, unambiguous cases in a single afternoon. Stumptails
reconcile at a high rate, within one or two minutes after a confrontation, and often with a lot of noise; you simply cannot
miss it!

After reading Alfred Brehm's unflattering description of the
bear-macaque, few people may be tempted to spend time
watching them. In a sense Brehm was right; at first sight stumptails do look, let us say, somewhat unusual. This is the reason
few zoos display the species. Anyone who knows stumptail
monkeys better, though, is smitten with their charming personality. Their looks were a running joke between my wife and the
Chinese researcher, RenMei Ren, who watched the stumptail
group for many hours a day: each time Catherine visited our
facility, she would accuse the monkeys of being ugly; RenMei
would literally jump up to defend them, saying "No, no, they
are beautiful!"

It is remarkably easy to tell individuals apart. The fur color
can be anything from gray, brown, or reddish, to black. The
face is covered with irregular patches and freckles in a pattern
unique to each individual. Also, the overall color and shape of
the face are highly variable. Of all the primates I know, including humans, stumptails are the species with the greatest individual differences in appearance. Some of the names for our
monkeys reflect this: the alpha female, Goldie, has a smooth
orange face and light brown fur; the second female, Wolf, has a

---

*The official name of this primate is stump-tailed macaque, but primatologists
usually abbreviate it to stumptail or even stump. The species is also known as the
bear-macaque.

Joey, a four-year-old male, has the freckled face characteristic of the
stump-tailed macaque. (Wisconsin Primate Center)

black face with prominent eyebrow ridges and a long-haired gray coat; the matriarch of the largest kinship unit, Silver, has a tomato red face full of wrinkles and, unique in the group, white fur. The coloration and build of Goldie, Wolf, and Silver are so different that one might think that subspecies have been mixed in captivity. This is not the case; a similar variation reportedly occurs within wild groups.

The female stumptail has a plump, pear-shaped body, whereas the male is muscular with broader shoulders. Only adult males are armed with long, sharp canine teeth. Both sexes are rather slow moving; they are built for walking, not climbing. Their expressive faces immediately capture attention because of the large size of the head relative to the rest of the body. It almost looks as if an ape head has been screwed onto a monkey body. This, and the virtual absence of a tail, is the reason why in the old days animal dealers could advertise stumptail monkeys as "pygmy chimpanzees."

In this chapter stumptails will be compared with rhesus monkeys, not apes. In terms of biological distance, stumptails are far removed from the line of evolution that produced humans and apes, but very close to the rhesus monkey and other members of the genus *Macaca*. The stumptail is a bit of an anomaly within this genus, but it is no doubt the best classification of the species. Interestingly, in spite of its close relation to the rhesus monkey, tremendous behavioral contrasts exist between the two macaques. I shall emphasize the contrasts at the expense of the similarities, as I wish to convey the diversity of social behavior.

Let me start with a popular view with which I do *not* agree. It is what Brehm called the stumptail's "phlegm." Originally this word referred to a body humor, which in medieval times was believed to cause sluggishness and apathy. Nowadays phlegm refers to such things as calmness and composure (in Continental languages the word is associated with the English gentleman). A similar opinion about stump-tailed monkeys was expressed by the psychiatrists Arthur Kling and J. Orbach, who compared them to—of all things—lobotomized rhesus monkeys. They observed a mild disposition, a natural docility, and the absence of malice. In the laboratory this is indeed often the case. Stumptails do not struggle or bite as much as rhesus monkeys, so caretakers can simply pick up the adult females instead of having to catch them by net. The females seem to realize that nothing is to be gained by resistance. But if something really important is at stake, their attitude can change dramatically— and then stumptails are more dangerous than rhesus monkeys.

We know little about their life in the wild, in Indochina and southern China, but several reports claim that male stumptails post themselves at the periphery of their group to warn others of danger and to protect them. This is not such a surprising strategy for a ground-dwelling species that lacks the speed for effective flight. Male stumptails are bigger and stronger than rhesus males and seem better coordinated. They are said to put up a fight when farmers chase them from the fields, and to attack human hunters to the point that some have never come back alive. In 1955, according to one unconfirmed report, the screams of a monkey shot by a hunter triggered a mass attack by an entire troop, in which the man was "ripped completely to pieces."

Mireille Bertrand, a French ethologist who observed stumptails in Thailand, once felt seriously threatened by the approach of a barking group. She believes that stumptails are prone to contagious aggression after gaining courage by a chorus of vocalizations. Against Bertrand herself they fortunately did not reach this point; she stayed motionless for twelve minutes, showing neither hostility nor fear. In captivity too, group-living adult males show fierce protectiveness. Their phlegm vanishes as soon as you try to catch one of the cream-colored infants. We have learned to be careful when we chase our group from the indoor to the outdoor pen. If one of the males refuses to move, standing his ground in a very determined manner, this means that we must have overlooked a youngster staying behind somewhere in a corner. We then leave the pen, allowing the juvenile to join the rest, followed by the male. Sometimes males even pick up the juvenile to carry him or her outside.

In summary, the picture of stumptails as lethargic and unlovely does not sit well with people who know them intimately. Both Brehm's description and the thought that a brain lesion might change a rhesus temperament into that of a stumptail are seen as insults to the species. Stumptail monkeys have strong, complex personalities of their own. They are highly intelligent, and although their temper is mild most of the time, they are full of life and energy.

Stumptail infants are cream colored for the first six months of life, after which their fur gradually turns dark. Compared to rhesus infants, they develop more slowly and remain dependent for a longer time. This fragile-looking infant is already four months old. (Wisconsin Primate Center)

Orgasmic Reconciliations

Species tend to attract the research that they deserve. We study aggression in rhesus monkeys, intelligence in chimpanzees, and song in gibbons. In the case of stumptails, the magnifying glass has been placed over their sex life. This is understandable in view of the incredible potency of this species, and the way sexual elements permeate group life from aggression to reconciliation.

Mephisto mating with Cinnamon. (Wisconsin Primate Center)

• It is quite normal for a male to copulate ten times in one day. The world champion is Sam, a male in a large captive colony, who once completed fifty-nine matings in six hours, each with an ejaculation.

• Both sexes may show the so-called orgasm face at the climax of intercourse; males show it virtually every time, females an average of once in six matings. With the lips pushed forward, leaving a rounded opening, the monkey utters a series of drawn-out vocalized expirations. The same behavior, together with body convulsions like those of an ejaculating male, may be seen in females when they embrace during an emotional reunion.

• After a heterosexual copulation the partners remain attached—almost like dogs, but capable of being separated if necessary—during which time the male is customarily harassed by a large number of other monkeys, who slap at him or pull at his hair, but who do not bother the female.

• Much of the sexual behavior of the species is independent of the female cycle, which lacks outward signs. Also, the seasons of the year have little or no influence on sex and reproduction.

• Males may bully a female with a mixture of aggressive and sexual behavior. They do so during tension in the group, especially when their position vis-à-vis other males needs to be demonstrated. Males bully strange females in the same manner. Bertrand introduced two females to a captive group and observed: "This forced mounting might be considered as rape, in the sense that the female was obviously unreceptive and unwilling. She kept crouching while the male forcibly lifted her hindquarters, shook and even bit her, and ignored her screams and dismount signals."

• Sexual elements are also prominent in reassurance and greeting behavior. Stumptails do not go as far in this as bonobos (treated in the next chapter), but their reconciliations are definitely "sexier" than those of most other primates.

Although my own work does not specifically concern sexual behavior, it is hard to avoid this topic when observing stumptail monkeys. In addition, because some of the most renowned specialists in this field are close colleagues of mine, I hear about it every day, so to speak. The above data come largely from Koos Slob and Kees Nieuwenhuijsen (the latter one of my former students at Arnhem) of Erasmus University in Rotterdam. Here at the Wisconsin Primate Center, the office next to mine

A puzzling phenomenon is the frequent harassment of copulating pairs by other members of the group. While Mephisto (*center*) and Silver are in the postcopulatory attachment phase, four adult females and one juvenile (*right*) rush to the scene. Mephisto gives an openmouthed threat to one of them, but can hardly defend himself until the tie is broken. (Wisconsin Primate Center)

belongs to David Goldfoot, who in collaboration with Slob and others was the first to demonstrate sexual climax in female primates, using stumptails as subjects.

Goldfoot's was a significant discovery, because until then many men of science—from Frank Beach to Desmond Morris, from David Barash to George Pugh—had assumed female orgasm to be uniquely human. While people readily accept the idea that male primates experience sexual pleasure, many become skeptical if the same is suggested for females. This orientation echoes the puritan belief, prevalent until the beginning of the century, that only men enjoy sex. Having left this erroneous opinion behind us, we still seem reluctant to view female orgasm as a widespread and natural phenomenon. To assume that female sexual arousal is limited to our own species is to deny it the deep biological roots of male arousal.

The official line of reasoning is that satisfaction is irrelevant for female primates; it is not necessary for copulation to occur, as males have appetite enough for two. According to the anthropologist Donald Symons in *The Evolution of Human Sexuality*, sexual enjoyment may even be dysfunctional for females if it makes them lose control over themselves. ("If orgasm were so rewarding an experience that it became an autonomic need, it might conceivably undermine a woman's effective management of sexuality.") Symons argues that successful reproduction requires that females carefully select their mates. Since this is less important for the male's reproduction, males can play the field. I believe that we should never place theory above observable facts. Female primates are equipped with a clitoris, an organ with only one known function. Furthermore, females are far from passive in sexual matters. They actively seek intercourse with males, and do so more often than is strictly necessary for reproduction. In the absence of physical reward this would be hard to understand. To me it would be the same as recognizing hunger as a driving force, at the same time doubting that it feels good to eat.

Goldfoot divided the orgasmic response into three parts: the subjective experience, the overt behavior, and the physiological changes. Obviously, only the last two aspects are measurable in female monkeys. He installed devices to gauge heart rates and contractions of the uterus and recorded the behavior on video. Using the Masters and Johnson criteria for humans, the monkey showed a sexual climax during mounts. At the very moment that the round-mouthed expression appeared on the female's face and the hoarse vocalizations were uttered, the equipment registered a sudden acceleration of her heart rate, from 186 to 210 beats per minute, and intense uterine contractions.

Actually, this experiment concerned reassurance behavior. The female's partners were other females. Female stumptails

Hold-bottom is a common conciliatory gesture. Dopey (*center*) teeth chatters as she clasps the hips of her presenting opponent (*right*). Yolinda (*left*) has been supported by Dopey in the preceding fight and now clings to her protector. Note the absence of eye contact. (Wisconsin Primate Center)

mount one another only during great agitation, which was created by placing together six females who normally lived apart. This caused a flurry of aggression followed by a series of mounts, as if some of the aggressive arousal had been transformed into sexual arousal. The mounting posture was the same as seen after fights in our social group. It is not a real mount, as when a male grasps a female's ankles with his feet, placing his hands on her shoulders. Instead, the pattern is what we call a *hold-bottom:* one partner pulls the other onto his or her lap, sitting behind and clasping the hips of the other. The resulting tandem position is the same as that of an attached male-female pair sitting down after copulation. It is a form of embrace from behind rather than a mount.

In short, it can be demonstrated that the sexual posture that stumptails often adopt during reconciliation is accompanied by physiological signs of orgasm. This is not to say that sexual climax is achieved during every reconciliation. One reason to doubt this is that pelvic thrusting and the orgasm face are uncommon during hold-bottom rituals in our large group. Instead, these contacts normally involve teeth chattering (a rapid opening and closing of the mouth with bared teeth), lipsmacking, or high-pitched squealing. The level of sexual arousal may be lower than that recorded in the experiment. Occasionally, however, we do notice an orgasm face during reconciliation. It is a revelation to know that nature has provided stumptails with a built-in incentive for making up with their enemies.

Two Macaques

Leaving her husband and (grown) children behind and taking a year's leave from Beijing University, RenMei Ren joined our team in 1984 to learn new techniques of behavioral research. She became absorbed by the family life of our stumptails and decided to replicate the reconciliation study we had previously conducted on the rhesus group. I had always wanted someone to do this, for preliminary observations had convinced me that stumptails provide a gold mine for peace research. To begin, we needed some simple comparison data on stumptail and rhesus monkeys. Lesleigh Luttrell and I collected hundreds of focal observations and conducted drinking tests on both species.

The groups are housed in adjacent, identical pens, but the

Mephisto, very protective of youngsters, occasionally gives them a ride. (Wisconsin Primate Center)

stumptail group is only half the size of the rhesus group. It includes two adult males, twelve adult females, one adolescent male, and a growing number of immature offspring. The oldest male is named Mephisto because of his devilish black face with bright red coloring around the eyes. He has a friendly disposition, though, and is popular in the group. Compared to Spickles and other rhesus leaders I know, Mephisto takes a more central position. While dominant rhesus males stay pretty much out of female affairs, Mephisto breaks up disputes among females and never fails to protect youngsters in need. Accordingly, he is an important refuge for attacked group members.

A close-range threat by Wolf (*right*) against Dopey. She stands her ground, staring back at Wolf. (Wisconsin Primate Center)

After major disputes Mephisto is always groomed by some of the antagonists, often of both sides. Everyone recognizes his influence.

Like all macaques, stumptails form matrilineal hierarchies. While their formal rank order (as expressed in submissive grins, teeth chattering, and other status communications) is as clearly recognized as that of rhesus monkeys, it is less strictly enforced. For example, if Goldie threatens Honey, a young adult female, Honey may bravely stand her ground, looking back at the alpha female. If Goldie threatens Dopey, one of the older females, Dopey may even threaten back. The result is a confrontation in which both rivals, faces close together, stare straight into each other's eyes with fierce expressions. Without breaking eye contact, Goldie may then grab Dopey's hand to give her a mock bite, pressing her open mouth on Dopey's wrist. This common gesture is unique for the species. It is not a real bite—although some investigators seem to record it as such—because it never results in even the slightest injury. Resistance to the symbolic punishment is rare. It happens regularly that a subordinate solves a tense standoff by *offering* her wrist for a ritual bite!

This would be a very stupid thing for a rhesus monkey to do. When a dominant rhesus female such as Orange threatens a subordinate, the first thing needed is distance. Staying close is dangerous; offering an extremity for biting is suicide. Counter-threats do occur among rhesus monkeys, but only from a safe distance—and certainly not against superdominants such as the alpha female. Most of their confrontations are totally lopsided, with one party being aggressive and the other submissive. Aggression is returned only one-third as often among rhesus monkeys as among stumptails. If, in a drinking test, a dominant rhesus monkey approaches the water hole with a threat face, subordinates obediently leave the hole 96 percent of the time. Rhesus monkeys punish unresponsive subordinates without mercy. Among stumptails, threatening dominants are avoided only half of the time. Many threats are ignored.

Sometimes I wonder how this is possible. Why threaten without backing it up with sanctions? What is the use of a threat that

is not taken seriously? Whatever the answers, relations among stumptails are rather egalitarian compared to those among rhesus monkeys. This is also appreciable in relaxed situations. For example, 70 percent of the more than ten thousand friendly approaches recorded in the rhesus group were made by the dominant party, and only 30 percent by the subordinate. Low-ranking monkeys are apprehensive; their passivity is an obvious way to avoid trouble. Subordinate stumptail monkeys, in contrast, are more self-confident, and initiatives for contact are almost equally divided between dominants and subordinates. All differences between the two species point at a loosening of the hierarchy among stumptail monkeys, whose group life is marked by considerable leniency and tolerance.

Against this background RenMei Ren conducted her study of reconciliation behavior. In the beginning the two of us watched together to make sure that the same criteria were used as in the rhesus study. Normally I prefer to tailor methods to the species under study. Both during play and during fights, stumptails do not run, jump, or climb very much. They shuffle around within reach of one another. It would have been logical, therefore, to abandon the definition of aggression used in the rhesus study, which required that one monkey pursue another for more than 2 meters. But this would have biased our comparison. A pursuit increases the distance between antagonists, which in turn affects the chances of a reunion afterward. To obtain fully comparable results, endless hours of waiting were invested to record enough instances of stumptails making the exceptional effort of chasing. We ended up with data on 670 pairs of opponents. Their behavior was recorded for ten minutes following the aggressive incident, and again on the next day as a control.

Using the same ten-minute observation window, we had pre-

A juvenile passively accepts a mock bite from an adult female. These bites are highly ritualized: they are always preceded by a threat and aimed at a wrist or a leg. Injuries never result. (Wisconsin Primate Center)

The physical closeness of stumptail society is visible in a wide range of situations. After having been rejected by his mother, a juvenile returns to her, teeth chattering at close range. (Wisconsin Primate Center)

viously found that rhesus monkeys seek contact with former opponents after an average of one in five conflicts. We found a much higher rate in the stumptails. They do so after one of two conflicts—56 percent of the time, to be exact. Reunions usually occur within one or two minutes and are markedly different from normal contacts. Most characteristic is the presentation of the rear end by one party, and a hold-bottom by the other.

Whereas the resulting tandem-sit is seen in fewer than 1 percent of the control contacts, it occurs in more than 20 percent of the reunions between former adversaries. Other typical behaviors are kissing (one monkey placing his or her lips on the other's mouth, often with some licking and smelling), teeth chattering, and genital inspection (mouthing, fingering, and smelling of the presented genital region). Our conclusion is that peacemaking among stumptails is both more common and more explicit than among rhesus monkeys.

So-called accidental contacts are not stumptail style. These macaques have little trouble looking each other in the face, whether they are high or low in rank, and do not need "excuses" to approach enemies. One of my students is now looking into another highly developed aspect: *public* peacemaking. Kim Bauers studies vocal expressions, ranging from the melodious cooing calls of lonely stumptail yearlings to ear-piercing screams during fights. Spectrography visualizes the structure of calls, and observation of the situation in which they are uttered helps to determine their meaning. It appears that special grunts announce an upcoming reconciliation, whereas loud squeals attract attention to the event itself. Only after major disputes do we hear these signals.

For example, the four adult daughters of Silver, and the matriarch herself, go after Yolinda, a low-ranking female. Mephisto protects Yolinda, making bouncing displays between her and the attackers; Honey, Yolinda's best friend, also gets involved. After screaming chaos the group calms down. Two minutes pass before Stella, one of the S-sisters, walks to a corner of the cage uttering a series of low-pitched grunts. Two of her sisters follow. Yolinda and Honey walk parallel to them in the same direction. Stella's grunts are joined by a chorus of the same sounds from the others. The grunting turns into high-pitched squealing when Yolinda presents to Stella. Hold-bottom is seen in several combinations of participants in the previous fight, both allies and opponents. At one point a hold-bottom "train" is formed among three females. Silver and other group members approach the scene, attracted by the commotion.

Such observations give the impression that stumptails have special sounds to make others aware of the restoration of peace—perhaps a unique trait of this species. Although little is known about their natural habitat, I always imagine these monkeys roaming the forest floor in dense foliage. In an environment where it is hard to keep track of events visually, it makes sense to inform the rest of the group vocally of important developments, such as the end of a dispute. In this way all who need to can get involved, and the mood of the group will settle.

The distinction between public and private reconciliation should not be confused with the distinction, treated in the previous chapter, between explicit and implicit reconciliation. While implicit reconciliations cannot be public, explicit ones can assuredly be kept secret. Children sometimes write little notes to their friends after a quarrel in the schoolyard, saying "I'm sorry" or "Let's forget it." These are explicit peace offers—albeit avoiding eye contact—but at the same time secret ones (unless the teacher intercepts one of the notes and reads it aloud, which is probably a good way to ruin the process).

In human societies public reconciliations involve public figures. In 1982 Harald Schumacher, goalkeeper of the West German soccer team, floored the popular French player Patrick Battiston with a sharp kick, which to many television viewers looked absolutely intentional. Battiston lost three teeth, had two broken ribs, and suffered a concussion; he stayed in the hospital for weeks. Reading French newspapers, one might think Germany and France were at war again. To calm the emotions, a press conference was arranged featuring both heroes. With a handshake, Battiston accepted Schumacher's apologies and both of them declared the collision an accident.

In recent years the media have brought us reconciliations of much grander proportions—or rather, steps toward reconciliation, because it would be naive to think that deep wounds heal after a single top-level meeting. There was the highly publicized gesture of French president François Mitterand, standing hand in hand with German chancellor Helmut Kohl over the graves at Verdun in 1984. Some journalists overemphasized the significance of this meeting, not mentioning that Verdun holds soldiers of the First World War, not the Second. Another historic meeting took place in Rome in 1986, when John Paul II was the first pope ever to visit a synagogue. He openly expressed disapproval of anti-Semitism, past and present, and called Jews the "brothers" of Christians. No doubt this was a significant step, although Jews felt that the mea culpa for the church's passivity during the Holocaust could have been more explicit. Forgiveness of large-scale atrocities is such a slow process that residues of antagonism completely disappear only when it hardly matters anymore. Thus, we are not surprised that the mayors of Carthage and Rome could come together amicably when they recently signed a peace treaty—2,131 years after the Roman army had destroyed Carthage.

The complexity and scope of such international events is, of course, not comparable with the described monkey reunions, but they have one principle in common: the reconciliation is signaled to the rest of the world. The stumptail's world is tight and small; ours is now about the size of the planet. Whatever the extent of the social network, it is essential for everyone to know in which direction enmities develop. Public peacemaking allows all parties, including those only marginally involved, to adjust their attitude; harmonization ripples well beyond the conflict's epicenter.

All-Embracing Unity

Stumptails resemble male chimpanzees in that they have turned the reconciliation process into a status ritual. Ninety-four percent of the time it is the subordinate who presents and the dominant who holds the bottom. Dominants may also pull at an arm or leg to deliver a mock bite—not with a threat face (we would count that as a new conflict), but a mock bite nonetheless. So the peacemaking process underlines who dominates whom. This is an interesting shift of emphasis compared to the rhesus monkey, who expresses dominance in the outcome of the fight itself rather than in the reconciliation afterward.

RenMei regarded presentation of the rear end as a formal

apology and the hold-bottom gesture as its acceptance. In her view, then, it is normally the subordinate who asks pardon. Not all reunions smoothly follow this pattern, however. By turning away, dominant monkeys may refuse to acknowledge a presentation. Even worse are instances of a dominant's trying

Mephisto (*right*) squeals while holding Wally's hips after a conflict among females in which the two males supported different sides. Wally utters low-pitched grunts. Months after this incident Wally became the dominant male. From then on the roles of the two males during hold-bottom rituals were reversed. (Wisconsin Primate Center)

to impose a hold-bottom on his or her rival. An uncooperative subordinate can get screaming mad if the dominant keeps pushing and pulling at his behind, thereby starting the conflict all over again, often at a higher level of escalation. The mechanism of conditional reassurance seems to be at work here. That is, relaxation of the relationship requires that both parties agree on their difference in rank. This rule applies most of the time, yet dominant individuals do make 6 percent of the presentations. We also know that they initiate over one-third of the reconciliations. Do dominants occasionally adopt the subordinate role to express good intentions? Is it because they regret their previous actions? Have they gone too far or been unreasonable?

In the same way that dominant stumptails have the flexibility to "apologize" to their victims on rare occasions that we do not yet understand, subordinates may take an aggressive role without automatically being perceived as challengers of the status quo. Let me give an example, which also provides a glimpse of this monkey's intelligence.

While the rest of the group huddles in the small indoor pen, Joey, the third-ranking male, and Honey sneak out for a rendezvous in the outdoor pen. Joey's sexual activities are not tolerated by the two older, dominant males. At the climax of their clandestine copulation, Joey shows the orgasm face and utters a grunt, the first of the drawn-out series of grunts that males customarily produce during ejaculation. Honey immediately turns her head to threaten Joey with a stare, whereupon the mating is concluded in silence. Perhaps Honey is concerned about Joey's noise, which might have given them away. This interpretation was supported a few days later when the two met again under similar circumstances. As they copulated, Honey turned before Joey had uttered any sounds, stared into his face, and briefly touched his mouth with her hand.

If Honey and Joey had been rhesus monkeys, things might have worked out differently. If a dominant rhesus female threatens the male she is mating with, he had better jump away immediately. If the female is lower in rank, on the other hand, her threat will have a confusing effect. Who would challenge a male in the midst of intercourse? I have never witnessed such a situation, but venture to guess that the male would break off the contact to chase the female. In any case, the female's threat would not be taken as a mere warning, as it was between Joey and Honey. Among rhesus monkeys menacing gestures are too much an expression of rank, and rank is too critical, to allow for such flexibility.

Unfortunately, our picture of primate group life has for a long time been determined by the example of rhesus society, as this was and is the most intensively studied species. We now realize that these infighters are by no means the standard primates we took them for, and that every species has developed its own variations on the themes of social organization. In addition to the variations discussed so far—the peculiar sex life, high social tolerance, and frequent reconciliations—two other features of stumptail behavior need to be mentioned.

In the first place, these monkeys are dedicated groomers. The average adult in our group spends 19 percent of his or her time grooming, compared to only 7 percent in the rhesus group. Champion among the stumptails is Cinnamon, who spends more than a third of her time in this meticulous labor.

Second, stumptails show a high rate of aggression. For adult rhesus monkeys I reported eighteen aggressive acts per ten hours per individual; for stumptails the average is thirty-eight. This figure may seem rather puzzling, given what we know about the species. The important point, though, is that the probability of escalation is extremely low. Only one out of every thousand confrontations leads to fierce biting, which makes the escalation rate eighteen times lower than that of the rhesus monkeys. The stumptail's high frequency of aggression is more than offset by the low intensity of that aggression, resulting in the rarity of violence.

The overall impression is one of extensive activity in all social domains. There is both a lot of grooming and a lot of squabbling going on. Stumptails continuously alternate between friend-

A row of huddled, grooming stump-tailed macaques. Social cohesiveness may be of vital importance for this species in the wild. (Wisconsin Primate Center)

liness and minor hostility, like an animated human family at the dinner table. Rhesus society is more disciplined: aggression is aggression, subordinates are well-behaved, and invisible fences separate the matrilines. All this is highly ambiguous in stump-tails. For example, reconciliation is not more typical of males than of females. It embraces the entire group, excluding no one. Unity and cohesiveness must be of paramount importance for this species. I suspect that in the wild they move and rest in close-knit groups and resolve internal conflicts with as little dispersal as possible. Perhaps they rely on communal male defense against predators; a cohesive group can be more effectively protected than one that is spread out.

How strongly unified male stumptails can be was dramatically demonstrated when our second male, Wally, had reached the age to challenge Mephisto. After a couple of fights, in which both males sustained minor injuries, Wally became the alpha male. The takeover took place within a few days and brought a reversal in sexual rights; Wally now copulated freely and openly, whereas Mephisto became more secretive. Once the matter of dominance was settled, the two males restrained their aggression in spite of lingering tensions. They still had many conflicts, during which Wally would fling himself at Mephisto. Since both males would bark and scream, the initial impression was of a very serious altercation. Yet at close range we could see that, instead of biting and struggling, the two rivals were using only their hands. They would mutually grab each other by the arms or shoulders, standing bipedally for less than a second before separating again. There were never any injuries. These strange sparring matches were always quickly reconciled, either with kissing or with a hold-bottom in which Wally, as the new dominant, held Mephisto's hips. The two males also groomed each other much more than usual. After several weeks their relationship relaxed. Wally and Mephisto acted in brotherly fashion again, and it became hard to tell who dominated whom, except during sexual competition or hold-bottom rituals.

As a consequence of his observations on three different macaque species in captivity, a French ethologist, Bernard Thierry, has started thinking along lines similar to our team. One of his species is the rare tonkeana monkey. This large black macaque—considered by many the handsomest member of the genus—is not a particularly close relative of the stumptail monkey but shares many of its behavioral characteristics. Thierry considers these traits to be components of a single complex. That is, the tonkeana monkey's low inclination to violence may be related to the richness of his reassurance behavior, the symmetry of his aggressive confrontations may reflect reduced fear of escalation, and the blurring of kinship lines may be caused by the permissiveness with which the young are raised.

Instead of considering one species as a prototype of how monkey groups are organized, we are beginning to see the incredible contrasts among species. Without exception, members of the genus *Macaca* establish clear-cut hierarchies, but the extent to which they follow their system is quite variable. Each species seems to strike its own balance between individual and collective interests, with "selfish" species emphasizing priority rights and "lenient" species sacrificing some of these rights for the sake of group unity and amicable relations. Field data are needed to understand the evolution of these differences in dominance style, and how they shape the overall group structure. It is an exciting task, one that eventually will illuminate our own societies.

CHAPTER FIVE

Bonobos

When did man emerge from the primates?
The question is really irrelevant. He was there
from the beginning.

> —John Napier

Numerous observations suggest that . . . the
female's purpose was to obtain food, not simply
to copulate. Presenting and copulation buffer the
food interaction and render the males more
tolerant.

> —Suehisa Kuroda

How would you feel if you were an African elephant
and knew nothing about your Asian counterpart, or if you were
a grizzly and knew nothing about the white bears of the Arctic?
Probably these animals could not care less, but we humans are
fascinated by our relatives and want to know them all. Yet the
bonobo, one of the four anthropoid apes, is still very much
unknown to us.

The most objective yardstick by which to measure the relatedness of different animals is analysis of the DNA molecules
that carry hereditary traits. This powerful new technique has
indicated a very close connection between species such as dog
and fox, and horse and zebra—nothing too surprising. The
shock came when an equally close resemblance, approximately
99 percent, was found between humans and two apes of the
genus *Pan*, the chimpanzee and the bonobo. Until the 1960s
science had followed the lead of Carl Linnaeus, who put the

human species in a separate category. The new data throw this two-century-old classification into question and show that Linnaeus' secret discomfort with it was justified. Later in life the Swedish naturalist came to regret his taxonomic decision. He had created the human slot, he said, to avoid problems with the church, in spite of not knowing any generic characteristic that sets humans apart from apes. Many people believe that the differences between themselves and the apes exceed 1 percent; but then, humanity does not have a history of unbiased judgment concerning its place in the universe.

It is estimated that the common ancestor of humans and the African apes lived about 8 million years ago. From an evolutionary perspective, with life on earth being 3.5 billion years old, it is as if the split occurred yesterday. The human-ape branch as a whole grew out of the primate tree about 30 million years ago. In other words, we share with chimpanzees and bonobos at least 20 million years of evolution that we do not share with monkeys. Small wonder that in every possible respect—anatomically, mentally, and socially—these apes differ more from monkeys than they differ from us.

DNA studies place the other two ape species, the gorilla and orangutan, at a greater distance from us. It appears that bonobos, chimpanzees, and humans are all more closely related to one another than any of them is to the two giant apes. This conclusion is still controversial, partly because its acceptance would spell the end of the old anthropocentric taxonomy. In anticipation of this moment it has already been suggested that the human race change its genus name from *Homo* to *Pan*, perhaps calling itself *Pan sapiens*, the wise chimpanzee. The alternative is to welcome at least two ape species to the genus *Homo*.

Because of their unique relation to us, it is a shame that bonobos are so little known. I feel I cannot describe bonobo behavior without first introducing the species. Where does the bonobo live? Why are all kinds of scientists suddenly interested? Discovery of this species, which has rightfully been called "one of the major faunistic events of the century," is bound to have as much impact on the way we look at ourselves

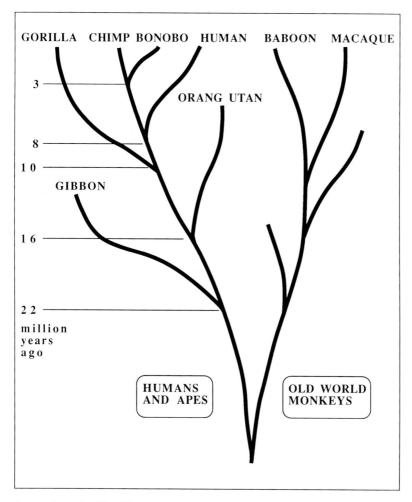

GORILLA CHIMP BONOBO HUMAN BABOON MACAQUE

ORANG UTAN

GIBBON

3

8

1 0

1 6

2 2

million
years
ago

HUMANS
AND APES

OLD WORLD
MONKEYS

Approximately 30 million years ago, the Old World primate line split into two branches, the monkeys and the hominoids. The second branch produced the common ancestor of humans and apes. The human line and the line of bonobos and chimpanzees divided an estimated 8 million years ago. (This evolutionary diagram is based on a comparison of DNA molecules by Charles Sibley and Jon Ahlquist.)

as did the much earlier discovery of the chimpanzee. The interpretation of our sex life particularly will never be the same again.

The "Pygmy Chimp" Is Neither

At the age of twenty-six Rembrandt van Rijn painted *The Anatomy Lesson,* in which Professor Nikolaas Tulp, surrounded by a group of attentive colleagues, dissects the left arm of a human cadaver. The instant fame that this remarkable work of art brought Rembrandt would probably not have been the same had the dead body been covered with fur. The same Professor Tulp in 1641 provided science with its first accurate description of an ape. He noted that some anatomic details so much resemble those of man "that you would scarcely see one egg more like another."

In that era European explorers were confused about the ape's place in nature. Gorillas, orangutans, chimpanzees, and the native people of newly discovered continents were barely distinguished one from another. This was not only the result of ignorance; there was also a certain unwillingness to make appropriate distinctions, that is, to accept non-Caucasians as full members of the human race. Until well into the last century, scientists continued to seriously compare "inferior types of mankind" with "superior types of monkeys." It was Western man's place in nature, not the ape's, that was at stake.

Tulp did not help to clarify things when he baptized his study object an Indian satyr, adding that the local people call it an "orang-outang." Instead of originating in the Indies, however, Tulp's specimen had come from Africa, probably Angola. Only its name came from the East Indies; in Malay *orang hutan* means man of the forest. Tulp's famous gravure of his satyr, redrawn over and over again in books of the seventeenth and eighteenth centuries, seems to show a female chimpanzee. At least this was the consensus until Vernon Reynolds, in 1967, boldly speculated that Tulp had dissected a bonobo.

Reynolds built his case on the small size of Tulp's animal ("its height resembling that of a boy of three years") and the webbing between the second and third toes. The joining of two digits of the foot is indeed characteristic of bonobos, but to recognize this detail in Tulp's illustration requires some imagination. Furthermore, in spite of the fact that the bonobo is also known as the pygmy chimpanzee, the species cannot be distinguished from the common chimpanzee on the basis of size. Common chimpanzees (classified as *Pan troglodytes*) are divided into three subspecies: adult males of the smallest subspecies weigh an average 43 kilograms, compared to 45 kilograms for male bonobos. Females of both species weigh approximately 33 kilograms. So Tulp must have dissected a juvenile, regardless of whether it was a bonobo (now classified as *Pan paniscus*) or a common chimpanzee.

I have overheard zoo visitors say "That must be the pygmy chimp!" while pointing at a two-year-old infant among the rest of the bonobo colony at the San Diego Zoo. Apparently these individuals did not consider the other apes "pygmy" enough to deserve the name. And rightly so; adult bonobos are big and strong. One day a zookeeper was lifted right off the floor when the adult male grabbed his arm with one hand through the bars of the night cage. Tulp perceptively wrote, "The joints are in truth so tight, with vast muscles attached to them, that he dares anything and can accomplish it." Again, his description applies equally to the bonobo and the chimpanzee.

What, then, are the differences? It is like comparing a Concorde with a Boeing 747, an orchid with a dahlia, a cheetah with a lion, or urban sophistication with country casual. I do not wish to offend any chimpanzees, but bonobos do have more style. The overlap in weight and size notwithstanding, all bonobos have a more gracile build than all chimpanzees. The bonobo's body is slim and slender. The head is smaller, on a thinner neck and narrower shoulders. The legs are longer and stretch while walking. The eyebrow ridges are thinner, the lips are reddish in a black face, the ears are smaller, and the nostrils are almost as wide as a gorilla's. Bonobos also have a flatter,

more open face with a higher forehead, and—to top it all off—
an attractive coiffure with long, fine, black hair so neatly parted
in the middle that you would swear each individual spends an
hour a day in front of the mirror. The quickest way to recognize
a bonobo is by this hairstyle and the light-colored lips. Bonobo
infants are even easier to recognize, because they are born with
a black face. Chimpanzees, in contrast, start life with a white
face, although many of them turn dark or even black after a
couple of years.

Strangely enough, the discovery of the bonobo took place in a
Belgian museum. It resulted from the close inspection of a skull
ascribed, because of its small size, to a juvenile chimpanzee. In
juveniles, however, the sutures between the skull bones ought
to be separated; in this exemplar they were totally fused. Obviously it had belonged to an adult with an unusually small head.
Ernst Schwarz drew the conclusion, and in 1929 described a
new subspecies of chimpanzee. In 1933 Harold Coolidge described the ape's anatomy in greater detail. He reclassified it as
an entirely new species, belonging in the genus *Pan,* the genus
that includes the chimpanzee.

Half a century after the discovery, sour grapes appeared on
the table. Coolidge said that he was the one who had first
noticed the fused bones in the museum. In his excitement he
had shown it to the museum director, who in turn had told
Schwarz two weeks later. Schwarz then grabbed pencil and
paper to publish the discovery. "I had been taxonomically
scooped," exclaimed Coolidge in a recent symposium.

To complete the historical overview, Amsterdam, the city in
which Tulp was perhaps the first to dissect a dead bonobo, may
also have had the first live exemplar on public display. In 1916
Anton Portielje wrote about a special chimpanzee, named
Mafuca, in Amsterdam Zoo. Portielje, a meticulous observer,

The bonobo has a more elegant build than the chimpanzee, and infants
are born with a dark face. (San Diego Zoo)

Between 1911 and 1916 two young apes, Mafuca (*left*) and Kees, lived at Amsterdam Zoo. Anton Portielje suspected they might be different species, yet it was not until 1929 that bonobos were officially recognized as a separate species. It is evident from this archival picture that Mafuca was a bonobo and Kees a chimpanzee. Mafuca was said to be the most popular animal at the zoo. (Photo courtesy of *Natura Artis Magistra*)

concluded that this ape represented "probably a new species." A photograph of Mafuca leaves no doubt that he was a bonobo.

Wild Bonobos and Wild Theories

Skulls and bones do not appeal to me. Thus the new ape did not exist, so to speak, until I first saw live bonobos in 1978. Since that day I have been looking for an opportunity to study the species, meanwhile collecting all the literature I could put my hands on. It is a very modest collection compared to the complete libraries written about chimpanzees or rhesus monkeys. Yet there is no lack of controversial claims. The bonobo has been called the most intelligent of all animals and the ape most resembling our ancestors.

Bonobos have a limited range in central Africa; they are found only in Zaire, south of the Zaire River. According to a recent field survey, there may be less than 100,000 members of the species. This number may sound reasonable, but forest destruction, ongoing in all tropical regions of the world, is seriously affecting the bonobo's habitat. Another threat is human predation: bonobos are hunted and eaten by the local people. Even in areas without bonobos, villagers mention their flesh as part of the menu.

Illegal sale is an additional factor. Foreign traders offer an amount equal to four times the local monthly wage for one young bonobo. Because the species is extremely susceptible to respiratory disease and pneumonia, very few captured animals survive. In 1959, when trade was still legal, all eighty-six bonobos ready in Kisangani for shipment to U.S. laboratories died in their cages in a matter of weeks. For every bonobo arriving in the West, countless others are lost. Clandestine individuals do still turn up, especially in Belgium, but fortunately nobody dares to buy them anymore. At the present time there are approximately fifty bonobos in the legal possession of laboratories and zoos. This is a small number compared to the thousands of captive chimpanzees.

Bonobos have been studied in their natural environment only since 1974. One project was started by Noel and Alison Badrian, a young couple of Irish and South African origin who had the courage and determination to enter the jungle on their own with almost no financial backing. The other project, at a different site, was started by Takayoshi Kano and a team of Japanese students. A decade of excellent research at both field stations, under difficult and isolated circumstances, has taught us how the social organization of the bonobo differs from that of the chimpanzee.

What the two *Pan* species have in common is a fluid social

structure: the members of a large community do not travel and forage all together; they form smaller bands, which change in composition through fission and fusion. The difference is that bands of adult male chimpanzees are quite common, but this is not true of bonobos. In bonobos the strongest attractions are among adult females and between the sexes; bonds among males are relatively weak. Hence the bonobo's society differs fundamentally from that of the chimpanzee, with female bonobos taking a much more central position.

Among male bonobos there are tensions, and sharing of food is rare. Females, in contrast, gather their food side by side. Highly prized items often are first claimed by the males but then shared with the females. Sex seems to play a crucial cohesive role in these foraging parties. Female bonobos are almost continuously receptive, that is, they brandish a pink genital swelling and are willing to mate during large portions of their monthly cycle. Not only do they mate with males, they also have sexual exchanges among themselves. Field-workers from both of the research teams have emphasized the role of sexual behavior in avoiding conflict, especially in relation to food. "Make love, not war" could be a bonobo slogan. These observations imply that the highly erotic life of the San Diego bonobo colony, which I will treat in detail, is not an aberration of captivity.

In *The Naked Ape* Desmond Morris presented humans as the sexiest of all primates. He argued that the almost continuous sexual receptivity of women is necessary to maintain pair-bonds. Pair-bonding, in turn, may be a way of avoiding competition among men. During hunting and territorial defense men need to cooperate closely; they cannot afford daily quarrels about sex. Tensions are reduced by dividing the women among them in a stable arrangement. Others have extended these theories. Owen Lovejoy, for example, brought bipedalism into the picture. Monogamous bonds allowed mothers to stay home with their progeny and take care of more than one dependent offspring at a time, which ape mothers cannot do. Because men therefore had to bring home food, walking on two legs became an important advantage: it left the hands free for carrying. In return for these male services, women made love with their caretakers.

As yet, all this is pure speculation—and many scientists disagree. Even a young *Australopithecus* female named Ruby has commented on the issue: "One thing hasn't changed in three million years. Males still think sex explains everything." This talking fossil, a product of the imagination of Adrienne Zihlman and Jerrold Lowenstein, also clarified her origin: " 'The little chimps were our ancestors, or so my grandmother told me. You have probably noticed how much I look like them' [the interviewers]. And so we had, but were too polite to comment on the truly striking resemblance." Ruby's public statements happen to agree with Zihlman's view regarding bonobos, the small chimpanzees. The bonobo's body proportions, especially its relatively heavy legs, are closer to those of *Australopithecus* than the proportions of any other living ape. Bonobos stand and walk on two legs more often, and with greater ease, than common chimpanzees, who do not straighten their backs as much. When bonobos stand upright, they look as if they had walked straight out of an artist's impression of prehistoric man. One conspicuous exception is obvious from the photograph on page 183: the feet of bonobos are not humanlike at all.

This picture was taken in captivity; it is not known whether bonobos in the wild use their bipedal abilities to any great extent. For the most part they knucklewalk* along the maze of paths covering the forest floor. Up in the trees they show a greater variety of locomotion, including walking on two legs. According to Randall Susman, bipedalism accounts for less than 10 percent of bonobo movement in the canopy, however. It is often used to try to intimidate elephants or unfamiliar field-workers from above. Bonobo scare tactics consist in branch waving, angry shouting, and urinating on the surprised targets. Interestingly, in view of the theories discussed above, another situation in which bonobos often walk upright is when they need both hands to transport large fruits.

---

*That is, walk and run quadrupedally on their feet and finger knuckles.

Bonobos show three of the elements present in scenarios of early human evolution:

(1) Females are sexually receptive for extended periods.
(2) Sexual life is rich and often related to food.
(3) Bonobos seem to walk on two legs more easily than other apes.

The remaining elements of the theories do not fit as well.

(4) There are no observations of striking cooperation among male bonobos.
(5) Female bonobos do not stay "home"; they move, with their offspring, over the same area as males.
(6) Pair-bonding is unknown for this species, yet males frequently accompany females, and these associations are said to be more stable and cohesive than in the chimpanzee.

Much still needs to be resolved, but it cannot be denied that the bonobo is a key species for the understanding of human evolution.

Let me add a rather wild speculation of my own. It concerns the bonobo's bipedal abilities and the sense of balance required for it. Bonobos are incredibly acrobatic. John MacKinnon observed them in the forest: "I had been amazed by the agility of bonobos; their treetop grace and sure-footedness were very different from the rather cautious and deliberate movements of chimpanzees." One may regard this agility, as MacKinnon does, as a sign that the bonobo is specialized for living in the trees. Yet there is no agreement on whether bonobos are more or less arboreal than chimpanzees. A completely different theory, first proposed to explain human bipedalism, may apply to the bonobo. Alister Hardy published it in 1960 after strong hesi-

Louise (*at left*) and Kevin on the lookout for the keepers. (San Diego Zoo)

tation: in his words, "it seemed perhaps too fantastic." Newspaper headlines casually summarized the theory as "Man Is a Sea Ape" and "Man Is Descended from a Dolphin," while it inspired Elaine Morgan to write *The Aquatic Ape.* According to this theory, our ancestors began walking on two legs while wading in the shallow waters of the coast, digging for shellfish and capturing fish. Walking upright gave an obvious advantage over groping about in the water on all fours.

Bonobos are remarkable in that they do not fear water. They surprised me tremendously by becoming playful on rainy days, wrestling and skidding about on the wet concrete. Frolicsome water ballets are unthinkable in chimpanzees, who hate rain. In fact, they have a special expression, called their rain face or dirty face, put on while they are sheltering from a downpour.

Bonobos have a remarkable sense of balance. Loretta calmly consumes some leaves while walking the rope. (San Diego Zoo)

With their lower lip stuck out and their upper teeth slightly bared, they are the picture of acute misery. Chimpanzees are known to drown in water, even when it is only knee-deep, because they cannot swim and panic quickly. As a result, zoos can keep them on islands. With monkeys this would not work—nor, perhaps, with bonobos. I do not know if bonobos actually swim (which would make them unique among the apes), but they are known voluntarily to enter pools or moats, splash in the water, even dive completely under.

The contrast between the two species may not be absolute (human-reared chimpanzees sometimes learn to like water, and not every bonobo is necessarily attracted to it), but in general the difference is striking. It is also understandable. The bonobo's natural environment abounds with rivers and streams, and swamp forests cover a large part of their range. Many areas are seasonally flooded. Although bonobos tend to prefer so-called terra firma forest, there may have been periods in their evolutionary past when water was even more plentiful than it is at present.

Both Kano and the Badrians have heard from local people that bonobos catch and eat fish. For many years field-workers found only ape footsteps and holes in the mud of small streams, but no direct evidence of fish catching. On a recent field trip, however, the Badrians saw two female bonobos walking upstream, in the water. They snatched handfuls of floating dead leaves, picking out things to eat. After the apes noticed them and fled, the investigators themselves tried the technique. They disturbed many small fishes hiding beneath the fallen leaves. Susman has observed that the numerous bonobo tracks along streambeds lack knuckle prints. This suggests to him that bonobos avoid getting their hands wet by assuming bipedal postures when crossing streams. So Hardy's aquatic ape theory, at least the part that links bipedalism to wading in shallow water, may help explain why bonobos have such strong, long legs.

Other elements of Hardy's theory are that water-dwelling primates—which, in his view, include humans—should show webbing between their toes (only 1 percent of humans do, however) and should grow long hair on their head both as protection against the sun and to give infants something to hang onto above the surface. Morgan further speculated that face-to-face copulation and a frontal vagina are human adaptations to an aquatic life-style. Other mammals living in this environment—dolphins, manatees, sea otters, and beavers—mate in the same way. The intriguing thing is that each of these elements applies to the bonobo. Webbing between the toes, although hardly noticeable, is common. The hair on the bonobo's head is longer than that of the chimpanzee. It is also a well-established fact that the bonobo female's sexual canal is ventrally directed, and that matings often occur in the "missionary position." A study published in 1954, before the sexual revolution, expressed it modestly in Latin: in the chimpanzee we see *copula more canum;* in the bonobo, *copula more hominum.*

In that the aquatic ape theory invokes images of protohominids with duck feet and goggles, it is hardly taken seriously by the scientific community. My application of the theory to bonobos is also written tongue in cheek. Still, I sincerely believe that the special relation of this species to water may have been more than marginally relevant during its evolution.

The Smartest Ape?

It is evident that the bonobo's smaller head and brain do not put him at an intellectual disadvantage compared to the chimpanzee. Confronted with a mirror, bonobos show all the signs of self-recognition; in captivity they are skilled tool users, and they exhibit high social intelligence. Some scientists even believe in their mental superiority. In 1985 the June 25 front page of the *New York Times* broke the news about Kanzi, a young bonobo male at the Language Research Center in Atlanta. According to Sue Savage-Rumbaugh, Kanzi learns to use geometric symbols representing words at a much faster rate than the two common chimpanzees trained earlier at the center. Kanzi also has a greater understanding of the spoken language of his teachers than ever before measured in an animal.

I have trouble with the claim that bonobos are the smartest of the apes. In San Diego I found their intelligence obvious and striking, and I have met Kanzi and agree that he is extremely talented. But it should be noted that Kanzi is the first ape trained in his mother's presence, which probably has a stabilizing effect on his personality. This may contribute to his performance. Also, many of the chimpanzees I know are far from stupid. The political manipulations of Yeroen, or the diplomatic skills of Mama, may have little to do with symbols or language, but they are no less impressive.

What should we think, then, of the often-quoted praise of Robert Yerkes for the intellectual capacities of Prince Chim? Yerkes in 1923 acquired two juvenile apes; the male was given the name Prince Chim and the female was called Panzee. Yerkes saw many contrasts between the two. His popular book, *Almost Human,* was inspired by Chim's unsurpassed "physical perfection, alertness, adaptability, and agreeableness of disposition." This anthropoid genius was contrasted to Panzee, whom Yerkes regarded as rather unintelligent. The primatologist doubted both apes were chimpanzees, but like Portielje before him, he lived in a time when the bonobo's separate status was not yet recognized. Only years after his death, when Harold Coolidge encountered his hide and skeleton in the American Museum of Natural History, was Chim declared a bonobo.

Yerkes' comparison of his two apes was not fair. In the first place, the common chimpanzee was suffering from tuberculosis, whereas Chim was healthy. Second, the more gracile build of the bonobo had led to an underestimation of his age. Chim was presented to Yerkes as less than two years old. Yerkes revised the estimate to more than three years of age, but Coolidge concluded from his postmortem inspection of Chim's dentition that he might have been as much as five or six years old. So the observed contrast in intelligence was between a distinctly older, vigorous bonobo and a terminally ill chimpanzee. Yerkes himself was well aware of the limitations of his comparison (in an article about the traits of young chimpanzees he commented, "It is clear that temperament and character are quite as dependent upon physical constitution as is intelligence"), but unfortunately his reservations are rarely cited.

Jacques Vauclair and Kim Bard compared the manipulative abilities of a human child, a chimpanzee, and the bonobo Kanzi. All three were seven months old at the beginning of the study. The child's object manipulations were the most complicated, but there were no significant differences between the two ape infants in this respect. One striking contrast, however, was that the child used her feet in 8 percent of the object manipulations, the chimpanzee in 7 percent, but Kanzi in more than 40 percent. In San Diego I noticed that bonobos use their feet completely interchangeably with their hands. They grasp food, kick at each other, and even gesture with their "handy" feet. In their equivalent of the chimpanzee's typical begging gesture (outstretched arm and open hand), bonobos may beg with outstretched leg.

There are many temperamental differences among the great apes. For example, Clemens Becker found that young bonobos completely dominated the play scene in a mixed group with young orangutans at the Cologne Zoo. Although much smaller in size, their play was so rough that the orangutans rarely dared to initiate wrestling matches. On the other hand, they were better at playing with objects. Orangutans are patient creatures; they play in a more constructive manner (building towers, for instance) than bonobos. Eduard Tratz and Heinz Heck noted another characteristic—namely, that the bonobo is the most nervous and alert species of ape. About the bonobos of the Hellabrun Zoo in Germany they wrote: "The bonobo is an extraordinarily sensitive, gentle creature, far from the demoniacal *Urkraft* [primitive force] of the adult chimpanzee." The Hellabrun bonobos were lost during World War II. They died of sheer fright, alarmed by the tremendous noise of the town's bombardment. None of the zoo's numerous chimpanzees suffered the heart attacks that killed the bonobos.

The problem of drawing the line between temperament and intelligence is most apparent in the usual comparison between dogs and cats. Dog owners believe their pet to be the smarter, not realizing to what extent their opinion is biased by the dog's tail-wagging desire to please her master. Similarly, the ease with which bonobos relate to people, and their natural alertness, may give them an advantage in certain experimental settings, but the differences among the four anthropoid ape species cannot be expressed by a simple scale of intelligence, from high to low. Instead, I see a multidimensional tableau of personality characteristics, some of which (emotionalism, for example) interfere with the execution of intellectual tasks, whereas other characteristics (such as the ability to concentrate) contribute to it.

The Peanut Family

In 1959 police officers of Zaire's capital city arrested a woman who allegedly had shot and killed her boyfriend because of his affair with another woman. They confiscated a baby bonobo found at the suspect's home. About a year later this little ape arrived at the San Diego Zoo. He was given the name Kakowet, after the French *cacahouette,* because he looked about the size of a peanut. (Estimated to be two years old, he weighed only 6.5 kilograms.) It did not take long for Kakowet to win the heart of everyone at the zoo. Even the discoverer of the species, Ernst Schwarz, who had never before seen a live bonobo, visited to get acquainted. The story goes that while he stood contentedly with the baby ape in his arms, Schwarz was greeted by a woman who said, "So you're the man who named that funny little monkey." A shocking thing to say to someone so familiar with the distinction between monkeys and apes!

In 1962 Kakowet's lifelong mate-to-be arrived. He and Linda became the world's most productive bonobo couple. Because each of her newborn infants was taken away to the zoo nursery, Linda delivered at unusually short intervals. Normally, female apes give birth every four to six years, but Linda had ten children within fourteen years. Many of the newborns were shown to the world on Johnny Carson's television talk show. I must say that I abhor the veterinary practice of removing offspring from competent ape mothers. The San Diego Zoo has since changed its policies and now has several naturally raised bonobos (born to Linda's daughters). The first of these more fortunate infants was two years old when I came to the zoo to do my study. Kakowet was no longer alive, and Linda and two of her daughters had been sent off to Atlanta on a breeding loan.

Colleagues in Wisconsin jokingly questioned the purity of my motives in going south in the middle of the snow-shoveling season. As it turned out, the winter of 1983–84 was one of the continent's worst ever. In southern California, however, I spent a virtually rainless (and snowless!) period with agreeable outside temperatures. San Diego has a beautiful zoological garden, not only because of the fauna, but especially because of the exotic flora covering the hilly terrain. Each day I entered this paradise, loaded with equipment and a large sign politely requesting the public not to disturb the observer or the apes. I conducted my observations while standing in front of the enclosure, separated from the crowd by a rope with a string of bright red flags.

At the beginning of my study the zoo kept its ten bonobos in three different groups. Two adult groups were housed in an old-fashioned "grotto"-type enclosure, with a dry moat separating the apes from the public. The groups were on display alternately: one day one group, the next day the other. This practice confused some regular zoo visitors. Once a little boy excitedly explained to his father how fast apes grow. He pointed at Kalind, a bonobo about his own age (seven), who was the smallest of his group. "In one week, in one week!" he explained, holding his hands apart to indicate the size of the two-year-old infant of the second group.

The four members of the third group were all under six years of age and seemed to miss contact with adults. They lived in an ideal huge enclosure with live trees to climb. Up in the trees and on the climbing structure the juveniles behaved quite normally. But as soon as they descended to roam the grass-covered ground they formed "trains," leaning with their arms and head on the back of the individual walking in front of them. The leaning was mostly done by the younger members, perhaps as a substitute for being carried on the mother's back. The way they clung to each other reminded me of the so-called together-together monkeys of Harry Harlow; rhesus monkeys raised in groups without their mothers get almost addicted to the comforting contact of tight huddles with their peers.

Although there was no lack of squabbles in the group of young bonobos, they were generally gay and frisky. In no way did they follow the pattern of cruelty and terror so powerfully

The juvenile subgroup at San Diego Zoo. *Left to right:* Lana, Akili, Kako, and Leslie.

Leslie biting Akili's finger. (San Diego Zoo)

described for a group of stranded human juveniles in *Lord of the Flies*. With this 1954 novel, William Golding did in literary fashion what Lorenz and other ethologists later did scientifically. The story called attention to the violent side of human nature. The message, although valid, was grossly exaggerated: the children grew bloodthirstier with every page. Meanness is common, among both human and ape children; but it is not totally unchecked, not even in the absence of adult supervision.

Once the oldest and most dominant juvenile, Leslie, was in a particularly nasty mood. She started the day scaring Kako, the youngest, by repeatedly chasing him out of the climbing frame for no apparent reason. She simply did not let him join the others, thereby upsetting the second female, Lana, who acted as mother and protector to Kako. Lana's protest barks did not help; they only led to a threatening charge in her direction as well. Leslie's next target was Akili, normally her chum. Akili was pursued from tree to tree and even briefly bitten. By the end of the morning all play had ceased and everyone looked nervous.

Eventually Leslie sat down a few meters behind Akili and Lana. I could feel the tension. The two looked around uneasily at her, then exchanged a glance between themselves. They must have agreed on what had to be done, because they rose simultaneously and together approached the boss to groom her. After approximately ten minutes Leslie relaxed, lying stretched out on her belly on a log, dangling her feet rhythmically beneath her. Akili, lacking the patience for long grooming sessions, left; Lana kept on. Later Leslie herself groomed Lana. There were no further problems that day.

Games Bonobos Play

The games of animals tell us something about their intellectual level, sense of humor, and temperament. The bonobos at San Diego Zoo are extremely playful and active, often romping about with the typical openmouthed play face. I learned a lot about the personalities of this fun-loving colony from the caretakers, especially Gale Foland and Mike Hammond. They know the apes intimately and are on excellent terms with them. On my first day Mike took me downstairs to tour the night cages under the outside enclosure. Though I stayed only ten minutes or so, the bonobos immediately recognized me among the crowd when I started observations three days later, wearing different clothes. This is all the more remarkable when we realize that these animals see 3 million human faces per year. Loretta turned her genital swelling to me, staring at me from between her legs. In reaction, Kalind—with all his hair on end—threatened me with grunts and an upswinging arm gesture. He seemed jealous of my receiving this sexual invitation. Although the message from the colony was mixed, I felt it to be basically a welcome.

Kalind. (San Diego Zoo)

The 2-meter-deep dry moat in front of the enclosure was made accessible to the bonobos by a chain hanging down into it. They could freely descend and climb up again. That is, unless Kalind tried to be funny. He loved to pull up the chain after someone, preferably the dominant male, had climbed down. Kalind would look down at Vernon with a play face and teasingly slap the side of the moat. On several occasions Loretta rushed to the scene to "rescue" her mate by dropping the chain back down. I believe these interactions must be based on empathy; that is, bonobos must be able to picture themselves in another's situation. Both Kalind and Loretta knew what purpose the chain served for someone at the bottom of the moat and acted accordingly; the one by teasing, the other by assisting the dependent party.

There were two games in which I was particularly interested, for they seemed to reflect higher mental processes. I have to be a little vague here, because obviously I cannot know what goes on inside the head of a bonobo. Yet I feel that some level of consciousness was involved. Both games were played daily in the juvenile group.

Blindman's buff. The bonobo covers his or her eyes either with an object (such as a banana leaf or a bag), by sticking two fingers in the eyes, or by placing an arm over the face. Thus handicapped, the ape stumbles around on the climbing frame, more than 5 meters above the ground. The game is played very seriously; I have regularly seen the bonobos almost lose balance, or bump into each other. Leslie is particularly good at it. She once demonstrated her familiarity with the frame by grabbing one of the ropes and swinging freely on it, landing and jumping off at the appropriate places while holding her eyes shut with one hand.

This game shows that bonobos are capable of self-imposed rules; it is as if they tell themselves, "I'm not allowed to look unless I lose my balance." Thus, they play with their perception of the world. The same game occurs in other apes and also in some monkey species, but I have never seen it played with such dedication and concentration as by the bonobos. The game is of

Lana playing blindman's buff with a banana leaf. *Below,* lacking an object to cover her eyes, she holds them closed with thumb and index finger of one hand and feels around with the other. (San Diego Zoo)

A sample of the funny faces made by juvenile bonobos. The expressions are complicated and hard to imitate. For example, Kako grins (*top right*) without retracting the corners of his mouth, and Lana (*bottom right*) makes quite a detour to get her fingers in her mouth. (San Diego Zoo)

course common in human children as well. Emily Hahn described it for both her pet gibbon and her daughter, Carol. The child, at least, could give some sort of explanation; she saw it as an existentialistic game. "There is no Carol," she said as she bumped into the dresser.

Funny faces. Juvenile bonobos often make weird facial expressions, aimed at no one in particular. To accentuate an expression, they may poke a finger into their cheek or stick out their tongue. Unusual jaw movements may occur; Lana often covers her upper teeth with her lip, bares her lower teeth, and pushes out her lower jaw while moving it rapidly up and down. This face pulling does not resemble the common communication of the species. Funny faces may be made while sitting alone, or while playfully tickling one another. For example, Leslie sits down above Akili, tickling him in the neck with her feet. Akili responds at first with a play face, then proceeds to fantasy faces with a bulging upper lip. Leslie too engages in facial acrobatics, doing her famous sucked-in face. The two cannot see each other's expressions.

Although young chimpanzees occasionally flip back their lips or put on other queer expressions, I have never seen primates in which the making of faces has developed into a sort of solitary pantomime, as it has for juvenile bonobos. Their striking voluntary control over the facial musculature makes me wonder how much of their normal communication is feigned or suppressed. Another question is what bonobos feel. If you try for yourself to make a particular face—a broad smile or a deep frown, for instance—you can feel an echo of the emotions normally associated with that face. Even completely novel, nonexisting faces feed back on our emotions. Do bonobos undergo the same experience, and is it for this reason that they like the game so much?

Kama Sutra Primates

A new zookeeper, unfamiliar with sexual encounters of the bonobo kind, once accepted a kiss from Kevin. Was he taken aback when he felt Kevin's tongue in his mouth! The habit of French-kissing is one of the striking differences between this ape's impassioned eroticism and the somewhat boring, functional sex of the common chimpanzee. Chimpanzees show few variations in the act, and most of their adult sex is connected with reproduction. Bonobos, in contrast, perform every conceivable variation, as if following the Kama Sutra. Their sex life is largely divorced from reproduction, serving many other functions as well. One, I am sure, is pleasure, and another is the resolution of conflict and tension. Obviously, the latter is the function that interests me most, but let me first describe bonobo sexual patterns.

Imagine the astonished glances of zoo visitors who would overhear me recording observations such as "All penises out at the moment." I measured sexual arousal by recording every five minutes which males had an erection. Since bonobos can sheath their penis, nothing is visible most of the time. When the organ does appear, however, it is not only impressive in size, but its bright pink color makes it stand out against the dark fur. Males invite others by presenting with legs wide apart and back arched, often flicking the penis up and down—a powerful signal. The bonobo male's genitals are among the largest in the primate world. Certainly relative to body size (and probably absolutely as well), this ape's testicle size and erect penis length surpass those of the average human male, until recently believed to be the champion.

Jeremy Dahl, who recorded the menstrual cycles of Linda and two of her adult daughters at the Yerkes Primate Center, found that they were in a sexually attractive state, with swollen genitals, almost 75 percent of the time. In the chimpanzee the figure is only 50 percent. Furthermore, other than in the pink swelling phase matings are virtually absent among chimpanzees, but occur regularly among bonobos. So the menstrual cycle imposes few constraints on the sexual behavior of bonobos—as is true of humans, with our even greater independence of the cycle.

Because the bonobo's vaginal opening and clitoris are frontally directed, face-to-face sexual postures are attractive and easy to achieve. In six years I never witnessed a frontal copulation in the Arnhem colony, except on one occasion when two chimpanzees mated through the bars between their night cages. In San Diego over 80 percent of the copulations between adult or adolescent bonobos of the opposite sex are face to face. The figure reported for the species in the natural habitat is around 30 percent. The generally higher frequency of frontal matings in captive studies is probably because it is more convenient for the female to lie on her back on a cage floor, or in the grass of an outside enclosure, than on a branch high in a tree. In fact, the number of frontal matings in the wild may be higher than reported because field-workers are rarely able to observe sex at the ground level; bonobos flee into trees as soon as they detect unfamiliar people.

Mating "doggy style" is somewhat problematic for bonobos because of the frontal vagina. Instead of lying flat on her belly like a chimpanzee female, the female has to lift her abdomen off

Face-to-face matings, here between Vernon and Loretta, are common in the San Diego bonobo colony.

the ground to allow the male to enter. Obviously, this is only a minor problem for a species that does not shy away from acrobatic sex, even to the point of mating while suspended in the ropes.

Copulations are short by human standards: they average 13 seconds, with a maximum of half a minute. Since the partners often have eye contact, the intercourse appears more intense and intimate than for species who mate from behind. In a frontal position, partners can better read each other's emotions. At the climax of the act, when the male slows down for his final, deeper thrusts, the female may bare her teeth in a wide grin, uttering some hoarse squeals. A similar sound was made by baby Lenore during her sexual contacts with Vernon. His penis being huge compared to the infant, the male did not even try to achieve intromission. Instead, he would take her on his belly and rub his penis against her fur. Or she would pull at his penis to make it appear, and then briefly press her vulva against it, demonstrating that she knew the connection. Lenore experimented in the same manner with the adolescent males.

Of the six hundred observed mounts and matings, less than two hundred involved sexually mature individuals of the opposite sex. In addition to sex with the infant, there were lots of "homosexual" contacts, especially after the two adult groups had been unified. Among males the patterns ranged from cursory mounts from behind to excited face-to-face embraces with thrusting and mutual penis rubbing. Although the males in these situations never achieved either intromission or ejaculation, their sexual contacts were uncommonly intensive for nonhuman primate males.

The two mature females, Loretta and Louise, engaged in a coupling pattern unique for the species. Known as GG-rubbing (genito-genital rubbing), it has been observed both in the wild and in other captive colonies. Bellies touching and faces close together, the two females rub their genital swellings together with rapid sideways movements. Sometimes one female lies on her back, but normally she clings to her partner, legs around the other's waist. The female in the upper position then lifts the

Both in the wild and in captivity, female bonobos are known to engage in intensive sexual contacts among themselves. Here Loretta clings with arms and legs to Louise in the typical GG-rubbing posture (*above*). In a variation (*below*) Louise lies on her back; Loretta rubs her swelling against Louise's, as Louise's infant looks on. *At right,* Louise (standing) invites Loretta to sexual contact. (San Diego Zoo)

other off the ground, as if carrying a child. It was almost always Louise, the older and more dominant of the two, who got the heavy job. On other occasions the two females would rub their swellings and clitorides together while facing in opposite directions—the one lying on her back, the other standing quadrupedally. Video analysis revealed that they stimulated each other with exactly the same rhythm of movement as that of a thrusting male: 2.2 moves per second.

The strong sensuality of the species was also evident from the frequent self-stimulation of lips, nipples, or genitals. If Kalind was frustrated because nobody wanted to share food with him, he would walk around with sulkily pouted lips, caressing one of his nipples with his thumb. Masturbation was performed with either a hand or a foot, but was never carried through to climax. That is, not by the males; the females were harder to judge. Males also stimulated each other by taking the shaft of the other's penis in their hand and gently moving it up and down a couple of times.

Among the juveniles, playful wrestling and tickling could develop into erotic games. When one of the males had an erection during the roughhousing, he might walk up to a play partner and stick his penis into the partner's mouth. Occasionally all four juveniles participated in group sex; some engaged in fellatio, while others happily wriggled their genitals together or tongue-kissed.

I never, at any moment, felt that I was watching a collection of pathologically oversexed animals. The sexual patterns described are by no means unique to this particular colony; they are characteristic of the species. The bonobos seemed to do what came naturally to them, inhibited only by the occasional jealousy of third parties (Vernon regularly interrupted sexual advances of the young males toward adult females). It would also be a distortion to view their behavior as an expression of "sexual freedom." Sexual morality simply does not enter the picture for bonobos, and in humans it is a highly variable item. The Victorian attitude toward oral sex, for example, is not universal. The rich, albeit apocryphal, history of this practice includes an ancient Chinese custom whereby grandmothers, mothers, and nursemaids calmed boy babies with fellatio, and the story that Cleopatra was known as the best "fellatrix" of her time.

We live today in societies that suppress sexual activities among children, with children, and among members of the same sex. The often-heard refrain is that since sex is "intended" for reproduction it should not occur outside that context. Even the pill is seen by some as an immoral device, in that it allows for nonreproductive sex. Insofar as these moral standards are based on what is supposed to be natural, they have no foundation in fact. Most animals engage in sexual activities at an early age, and I would not be surprised if many problems with sexual obsessions and frustrations in our own societies result from the guilt with which we surround such experimentation and rehearsal play. Also, intercourse between partners of the same sex is not at all unusual in animals. What *is* unusual is an exclusive orientation to same-sex partners. The high frequency of such orientation in humans is not yet understood, but here again intolerance may play a role. By forcing a choice, it leads to a sharper division than would otherwise be necessary between those who are homosexually oriented and those who are not.

A more serious matter is that concerning sexual relations of adults with children. There are those who employ an overly broad definition of sexual abuse of children, including almost any form of sexual behavior, some of which cannot be suppressed without infringement upon forms of body contact that many consider perfectly normal. If a mother cleaning her child kisses and touches it everywhere *except* on the genitals, this cannot but cause a vague awareness in the child that there is something very wrong with these body parts. The main distinction I see is not between contacts that do or do not involve the sex organs, but between loving and responsible relations of adults with children, and relations sought purely for the adults' own gratification. The last type of relationship can easily become abusive—that is, harmful to the child—especially if combined with physical or psychological coercion.

I once had the opportunity to present my observations on bonobos at a meeting of experts on child abuse. They agreed that although Vernon, Kevin, and Kalind might be sentenced to twenty years in jail if they were members of a Western society, their behavior had no really worrisome elements. The males never mounted juveniles and infants without their "consent"—anything else would have been evident from the youngsters' struggling and the males' restraining attempts. The contacts were brief, friendly, often sought by the youngsters themselves, and without penetration. It may well be that sexual abuse of children is a uniquely human pathology.

Bonobos usually do not have sex just for fun. As explained below, the majority of mounts and matings occur in tense situations. Both the romantic notion of sex as "making love" and the conception-oriented view of sex should, as far as these apes are concerned, be extended to the concept of sex as an alternative to hostility. Bonobos use sex in the same way Chinese grandmothers are said to have used it, as a pacification. Thus, erotic interactions among bonobos may be essential for group harmony. This broader function does not conflict with the reproductive function, as males do not ejaculate during contacts with partners other than mature females.

The Sex-Contract Hypothesis

At around noon each day the apes received an extra amount of food—either a bundle of branches and leaves or some juicy sprouts of banana trees. Penile erections occurred as soon as the long-awaited keepers came in sight, reaching a peak when the browse was thrown into the enclosure. Normally, the males had erections less than 5 percent of the time, but at feeding time the figure increased to over 50 percent. This simple statistic illustrates the link between food and sex, which is *very* close in the bonobo. Sexual behavior is an integrated part of the species' begging and food-sharing behavior.

Initially, food is claimed by the most dominant individual of

Kako, with a pout face, touches the food of his best friend, Lana (*above*).
Minutes later, Lana (*right*) and Kako embrace in mutual defense against
a charge by Akili (*left*), who does not have any food. (San Diego Zoo)

the group, with the others gathering around for their share. They rarely beg with hand gestures; instead, they put their face close to the dominant's face, following intently the consumption of each and every leaf. They may get so absorbed in the spectacle that they start mimicking the dominant's chewing movements. This, together with the chorus of excited grunts and peeps, creates the impression that everybody is enjoying the meal even before distribution of the food has taken place. Subordinates also express their desire by uttering soft whimpers and adopting an expression known as the pout face. This pitiful signal is used especially when dominants refuse to let go of a leaf eagerly touched and smelled by one of the beggars.

Bonobos, with their well-controlled emotions, rarely throw temper tantrums when frustrated. Only once did I see an adult come close to a tantrum. Loretta was not in the swelling phase, which made it harder for her to obtain food from Vernon. After having followed the male with his huge bundle of branches for a while, she saw him reach for the chain to the dry moat. She embraced him with a pout face, and when this failed to stop him, she threw herself at his feet with a few spasmodic body movements. Compared to the din of chimpanzees under comparable circumstances, Loretta's display of despair was brief and civilized. It also did not help. It should be added, though, that Loretta always ended up with most of the food, whether she had a genital swelling or not. The only variable was the length of time Vernon maintained possession, which never exceeded ten minutes.

In the second subgroup the situation was different; Louise was a selfish dominant. Kevin had to wait a long time before he could join the meal. For example, one day four heavy chunks of banana tree were provided. After Louise had claimed the first two pieces, Kevin approached the third. He carefully touched it, then licked his fingers, all the while glancing at Louise. When he finally dared to take the chunk, Louise lunged at him, appropriating the food. She also claimed the fourth piece for herself. Because her eyes were always bigger than her stomach, it was no surprise that two untouched pieces remained for Kevin half an hour later.

With less heavy food items, Kevin benefited from his popularity with the infant, Lenore, who had free access to her mother's food pile. Once Louise took the entire bundle of leaves thrown into the enclosure. Penis erect, Kevin invited baby Lenore. She climbed onto his belly and he made thrusting movements. Then Lenore nonchalantly took two branches from Louise's heap and pushed them over to Kevin. The male patted the infant kindly on the back with one hand, while picking up the branches with the other. He installed himself at a distance from Louise, eating the leaves brought by Lenore's courier service. Lenore was not always successful, as her mother knew what she was up to; there was no problem if Lenore fed on the browse, but as soon as she started to make off with a branch, Louise tried to retrieve it.

Food distribution among apes seldom involves active giving. The following striking exception was observed among the San Diego bonobos by Thomas Patterson in 1971. Linda's two-year-old daughter whimpered, looking up at her mother with a pout face. Linda did not have any food but seemed to understand what the infant wanted. Normally, infants give this signal when they want to be nursed, but remember that all of Linda's offspring were human reared. The infant had been returned to the group long after Linda stopped lactating. Linda went to the fountain to suck water into her mouth. Then she sat down in front of the infant, leaning forward with puckered lips. The infant drank from her mouth. Linda repeated her trip to the fountain three times.

This is a telling, yet uncommon example of sharing. During my study, the "flow" of food from one individual to another was effectuated mostly through one of three patterns: begging, relaxed cofeeding, or assertive claiming.

Begging sometimes was rewarded with a substantial portion of the possessor's food. If Vernon, for instance, was approached many times by Loretta and Kalind, he would break his bundle of branches in two and walk off with the smaller part. This came close to an act of giving: Vernon was of course not unaware of the fate of the food he left behind.

Cofeeding on a shared pile of food was typical between very intimate individuals, such as Louise and her daughter or Lana and her adopted son, Kako. This most tolerant pattern was also common between the two adult females.

Obviously, assertive claiming of another ape's food was the method of dominants. But subordinates also showed this behavior, especially Loretta when she was in her sexually attractive state. She would mate with Vernon and make high-pitched food sounds afterward, while taking over his bundle of food. She would hardly give him the opportunity to pull out a branch for himself. Sometimes she grabbed the food out of his hands in the midst of intercourse. People watching my videotapes of these scenes cannot resist drawing a parallel with prostitution; but there are problems with this interpretation.

The loaded term "prostitution" was first used in connection with nonhuman primates in the thirties, by both Solly Zuckerman and Robert Yerkes. Yerkes concluded in 1941 that "female ability to trade on the sexual relation is incomparably greater than that of the male." His interpretations were mocked by Ruth Herschberger in *Adam's Rib*. Herschberger let one of Yerkes' female chimpanzees, Josie, take a stand against being described as the "naturally subordinate sex" to which the other sex may temporarily defer, allowing her to act "as if dominant." Josie complained: "Can't I get any satisfaction out of life that isn't *allowed* me by some male chimp? Damn it!" This criticism has recently been echoed by other feminist authors.

Let us first get away from a term tainted by associations with dark street corners and obscure "establishments." Helen Fisher, an anthropologist, in 1983 wrote *The Sex Contract*, which propounds a theory inspired by the ideas of Owen Lovejoy and Desmond Morris. Female protohominids, the theory runs, used sex to establish stable relations with males in order to benefit from their care. Fisher sums up: "Males and females were learning to divide their labors, to exchange meat and vegetables, to share their daily catch. Constant sex had begun to tie them to one another and economic dependence was tightening the knot." The key to such an arrangement, according to Fisher, was the female's strong sex drive, her capacity to mate throughout the cycle, and a preference for face-to-face copulation (which nurtures intimacy, communication, and understanding). Because the author believes these developments to be uniquely human, she must be ignorant of the behavior of the bonobo, to which large parts of her theory seem applicable.

If the sex-contract hypothesis disturbs feminists, this is because it proposes a distinctly different power base for males and females. The male base is physical dominance; the female base, sex appeal and family ties. The one form of power is not more "natural" than the other, but the female's power is less straightforward. Nobody who has seen a male chimpanzee fight a female can doubt that females need resources other than muscles and teeth to achieve their goals. Thus, they fight their battles with the opposite sex in other social arenas. To what extent this provides them with influential positions decidedly depends on male temperament; subtle tactics are of little use in dealing with merciless bullies.

Most of the time, male primates are restrained in their aggression against females and youngsters. In many zoo colonies gorilla males dominate females only up to a point. If the male flaunts one of the unwritten rules of the group, the females gang up on him. An adult male gorilla is the most formidable fighting machine of the primate world, undoubtedly physically capable of holding off, even killing, a number of the much smaller females of his species. Psychologically, however, he appears incapable of fully exploiting this advantage. It is quite spectacular to see an alliance of barking females chase—even beat—the gargantuan male, whose hands seem tied behind his back by the neurons in his brain.

Vernon showed similar inhibitions, not only when he lived with both Louise and Loretta (who together formed a strong team, because bonobo females are much closer in size to males than gorilla females), but also when he lived with Loretta alone.

Skinny Loretta was no match for Vernon, a muscular male with full-size canines. Nevertheless, he avoided physical confrontations. Once, when Loretta caught the browse food in midair before Vernon arrived, he charged closely past her a number of times, with all his hair on end. Instead of being impressed and dropping her booty, Loretta flew into a rage, whistle-barking and gesticulating wildly at the troublemaker. Vernon then vented his feelings on Kalind, who fled.

If Vernon's gentleness with females is typical of male bonobos, this trait promotes a more egalitarian relation between the sexes than is found in many other primates. I know of several captive bonobo groups that are dominated by a female. The question is whether sexual intercourse also contributes to such a power balance. I think it does, but not exactly in the calculated, transactional fashion suggested by the theories treated above. If I had observed only Vernon and Loretta, I might have agreed, because they did seem to follow this pattern. However, the sex-contract hypothesis cannot account for all the sex that goes on among bonobos at feeding time. It occurs in every age and sex combination, whether food is shared or not, and it is initiated by dominants as well as by subordinates. There were, for instance, as many food-related matings between Louise and Kevin as between Loretta and Vernon. In the first pair, Louise was completely dominant and had all the food she wanted. After mating she rarely shared with Kevin. While this does confirm Yerkes' claim that the male benefits less from sexual contact than the female, it also demonstrates that the female does not necessarily have a "price" in mind.

Needless to say, there is more aggression at feeding time than at other times of the day. In contrast to the sex-contract hypothesis, which stresses the trading of favors, I emphasize the attenuation of competitive tendencies. For much of their rivalry bonobos substitute enjoyable erotic activities. Their sexual behavior is related to *tensions* over food, rather than to the food itself. This mechanism may be exploited by some individuals, such as Loretta, as bonobos are smart enough to learn how sex

In spite of many differences in the way chimpanzees and bonobos reconcile, both species use the outstretched hand to invite their adversaries. (San Diego Zoo)

mollifies dominants. Such tit for tat is a secondary development, however; conflict resolution is the more fundamental and pervasive function of bonobo sex.

To complicate matters, there is a third view, a very simple one—namely, that sex at feeding time is nothing more than an arousal phenomenon. According to this explanation, excitement over attractive food translates into sexual arousal. One of the purposes of my study was to resolve this issue. If, as argued above, the bonobo's sex life serves to smooth tense or disturbed relationships, the presence of food should not be a prerequisite; what is important is whether there is a potential for aggression. The critical question therefore becomes, Do these apes also use sex to resolve conflicts unrelated to food?

Sex for Peace

At the Japanese research station at Wamba in Zaire, bonobos are enticed out of the forest by daily provisioning of sugar cane. This procedure has allowed Suehisa Kuroda, Akio Mori, and others to investigate the link between food and sex. Mori compared his observations with those on common chimpanzees in the Mahale Mountains, where the same provisioning technique is used. He was struck by the sociability of bonobos and reached the same conclusion I did. Peaceful aggregation is made possible, according to Mori, "by changing the character of sexual behavior into affiliative behavior in which all individuals can participate, and by decreasing the reproductive meaning."

Since artificial feeding has the disadvantage of disturbing normal foraging habits, Western primatologists at a second field station work without this technique. The problem is to track and watch bonobos in the dense forest without scaring them away. In the rare event of an aggressive pursuit among the apes, the researchers are usually left to speculate on the outcome as the contestants disappear from sight. Owing to these problems, there are no field data on peacemaking among bonobos outside the feeding context.

At this point my own study becomes unique; I had no trouble following former adversaries for some time to see whether they reconciled. Although the frequency of reconciliation may be higher in an environment where apes cannot avoid one another, the *way* they make up is probably the same as in the wild. I returned from San Diego with more data than I had dreamed possible. It took my assistant, Katherine Offutt, and me more than a year to process the five thousand and more social interactions captured on video or described on audio. This material is now conveniently stored in the computer of the Wisconsin Primate Center. Let me explain the analysis most pertinent to our problem, which concerns several hundred hostile encounters unrelated to food.

The computer was programmed to compare behavior before and after each conflict with the so-called baseline level, that is, with behavior during the rest of the day. It was found that individuals groomed each other less than normally before an outbreak of aggression, and that grooming continued to stay below the baseline for some time afterward. The graph of embracing, friendly touching, and sexual contact, on the other hand, showed the opposite pattern. Immediately after an aggressive incident, the rate of these contacts jumped, staying well above the baseline for twenty-five minutes.

For some antagonists the measured change was quite dramatic. For example, Vernon regularly chased Kalind into the dry moat, mostly in reaction to the adolescent's sexual advances to Loretta. After such incidents the two males had almost ten times as many intensive contacts as was normal for them. Vernon would rub his scrotum against Kalind's buttocks, or Kalind would present his penis for masturbation. On other occasions they embraced, then indulged in some rough tickling games. Without such contact Vernon would not allow Kalind back into the enclosure. So after emerging from the moat, Kalind's first task was to hang around the boss and wait for a cordial signal.

Often Kalind was so fearful that he mistrusted even the friendliest overtures. If Vernon reached to hold the youngster's outstretched hand, Kalind would withdraw. Then Vernon would make the typical "come here" gesture—open hand and rapid finger movements toward himself. After much hesitation, Kalind would touch Vernon's hand, in a scene reminiscent of the two hands at the center of Michelangelo's *Creation of Adam*. Only when this brave touch went without unpleasant consequences would the younger male venture within Vernon's reach, screaming excitedly during the ensuing reconciliation.

In the other subgroup, an interesting episode once followed a quarrel between Kevin and baby Lenore. Kevin had been playing with some chains and ropes, arranging them in a circle around himself in the manner of wild apes building a nest of tree branches. While Kevin was away gathering more materials, Lenore sat in the empty nest. Upon his return, Kevin chased her out; but Louise heard her daughter's shrill peeps and barked a warning. This gave the infant instant courage, and she

A sequence of conflict and sexual reconciliation between Vernon and Kalind.

This page: Vernon charges at Kalind, who has disappeared from view. Minutes later, Kalind (*right*) approaches Vernon, making eye contact from a safe distance. His froglike posture indicates readiness to flee.

Next page: Vernon (*now on right*) embraces a nervously grinning Kalind; again note the eye contact. In the final scene Vernon gives Kalind a genital massage. (San Diego Zoo)

Vernon (*left*) punishes Kalind for his constant begging by rhythmically punching him in the chest. Kalind passively accepts the punches, then leaves Vernon alone. Such inhibited forms of aggression are typical of bonobos. (San Diego Zoo)

charged at Kevin. The confrontation went back and forth until Louise retrieved Lenore and calmed Kevin by putting an arm around his shoulder. Minutes later baby Lenore approached her adversary with a genital present, lying on her back with her legs apart and her vulva pointing at the male. Kevin mounted her, making thrusting movements, then carried her away. In so doing he made a grave mistake; it was the first and only time I saw him carry the infant out of her mother's sight. As soon as Louise realized what had happened, she rushed around the enclosure until she found the two. After collecting her pride and joy, she punished the offender by biting his toe. Soon, however, she returned, took Kevin's foot in her hand, and carefully licked his blood.

These are only a few instances of many; increased contact after aggression was a widespread phenomenon. While chimpanzees kiss and embrace during reconciliation but rarely engage in sex, bonobos use the same sexual repertoire then as at feeding time. This is the first solid evidence for sexual behavior as a mechanism to overcome aggression. Not that this function is absent in other animals (or in humans, for that matter), but the art of sexual reconciliation may well have reached its evolutionary climax in the bonobo.

With this in mind, many encounters take on a special meaning. Once, in the juvenile group, Leslie found Kako blocking her way on a branch. First she pushed him. Kako, who was not very confident in trees, did not move but tightened his grip, grinning nervously. Next Leslie gnawed on one of his hands, presumably to loosen it from the branch. Kako uttered a sharp peep—and stayed put. Then Leslie rubbed her vulva against his shoulder. This calmed Kako, and he walked out in front of her. It appeared that Leslie had been very close to using force, but instead had reassured both Kako and herself by means of genital rubbing.

In a different incident the two adult females, Louise and Loretta, engaged in a competitive game over a cardboard box. They were in a playful mood, romping around, grabbing the box away from each other. Louise tried to control the game. She resisted giving the box away and pummeled Loretta quite roughly if she persisted. All this was done in fun, with broad play faces and laughing sounds. Yet in young apes such games may turn into fights, and I had the feeling that the two females were approaching that point. Then a new element was introduced. When the pulling became tense, Louise invited Loretta for GG-rubbing. This happened several times during the game, and the unhappy ending never occurred.

It is uncertain whether the absence of aggression in these cases can be ascribed to the use of the sexual "alternative."

Behavior that *almost* happens cannot be measured. Ethologists need new techniques to record subtle changes in animal moods and intentions so that we can learn how conspecifics react to them. If sensitivity to human emotions is any measure, apes must be masters at the detection of nonverbal clues. Nobody working with adult anthropoids escapes the feeling of being uncannily transparent. Apes respond to all sorts of moods before we humans realize how nervous, depressed, or irritable we are that day. And they read our mind when we try to hide something disagreeable, such as an imminent visit by the veterinarian. Among themselves this perceptivity must allow apes to anticipate conflict situations just as we do, and to take preventive measures whenever possible.

Conflicts in the bonobo colony were more often reconciled than those in the chimpanzee colony of Arnhem Zoo. I hasten to add that it is difficult to evaluate this finding in view of the contrast in group size and available space of the two colonies. A second major difference, however, seems hard to ascribe to living conditions. Reunions of former bonobo opponents were initiated for the most part by dominants, which was not true of chimpanzee adversaries. Since aggressive confrontations are normally started by dominants, the finding implies that peace efforts among bonobos are typically made by the offending party—almost as if they regret having lost their temper. Thus, although the youngest, most helpless individuals received quite a number of threats and mild punishments, these were almost invariably followed by reassurance afterward. This was true of all subgroups of the colony. As a result, social life gave the impression of being ruled by *compassion.*

In this light it is not surprising that aggression among bonobos never involved the prolonged hitting, trampling, and biting occasionally seen among chimpanzees. Physical aggression was not absent, but it rarely lasted for more than a second. Kevin would charge past Lenore and slap her on the back, or Leslie would grab Akili's hand, bite his finger, and release him. Attackers might storm wildly forward, only to put on some inner brake before a collision could occur. Once Vernon kept chasing poor Kalind to the point of exhaustion. When the adolescent finally crouched in a corner, I expected a real dressing-down. Vernon, hair on end, made his final charge, halted just in time, then gently poked his screaming victim in the back and strolled away as if he had never intended to do anything more.

Although the remarkable gentleness of the bonobo species has been noticed by other investigators, we should also realize that, until a decade ago, the same opinion prevailed with regard to gorillas and chimpanzees. Now we know better. Recently, Takayoshi Kano published a rather shocking report on physical abnormalities in free-living bonobos. An astonishing number of them lack fingers, toes, even entire hands. Two-thirds of the males and one-third of the females show limb abnormalities. Kano produced a whole list of possible causes: congenital malformation, poacher snares, poisonous snakebites, falls from trees, and intraspecific violence. The higher incidence of missing or deformed body parts among males, especially adults, supports a link with aggression. And the tendency of bonobos to aim bites at the extremities may explain the nature of the defects.

Kano even encountered one male lacking both testicles, which could have been the result of a fight injury. This possibility reminds us of Luit's fate and warns against idealization of bonobos. They have not been studied nearly as intensively as the other apes, and little is known, for instance, about their intercommunity relations. If violence does occur, presumably it is chiefly during territorial encounters.

As yet, all of this is conjecture. But let me stress again that within the social group bonobos are considerably less belligerent than their closest relatives. Severely damaging fights have never been observed, not even when at Frankfurt Zoo a father and son went through a serious power struggle. The two males occasionally pummeled each other with their fists, but only two minor wounds were found on the older male during the stressful months in which he lost his position. Furthermore, the two were in a cage, which also housed four other individuals. Evidently conflict management among bonobos is so highly developed that de-escalation is the rule, and escalation the rare exception.

Epilogue

Several weeks before my departure the two adult subgroups at the San Diego Zoo were unified. Their members were far from strangers, as they had seen and heard one another every night in their sleeping quarters. They had not had physical contact for

Vernon (*left*) and Kevin during their tense introduction. Kevin spreads his legs to present his penis. Later the two males embrace and calm down. (San Diego Zoo)

two years, however, so that we were quite apprehensive, especially about Vernon and the adolescent male Kevin. Would they injure each other? To keep things better under control, we scheduled an initial introduction of just the two males. A team of people assisted me in recording all details, and the keepers were on the alert to open a trapdoor in case Kevin needed an escape.

First, the two moved around each other with an emotional dialogue of rapidly alternating screams. Both had erections, which they regularly presented to each other. Kevin invited Vernon with open-hand gestures, sometimes impatiently shaking his hand, sometimes beckoning with his fingers. The younger male did not dare to approach, however, and Vernon also kept his distance. Their screaming concert lasted a full six minutes; it did not sound aggressive at all, rather extremely nervous. It was as if each male wanted to make contact but did not know whether the other could be trusted. Suddenly Vernon responded to one of Kevin's sexual invitations by rushing toward him. They embraced frontally, with broad grins on their faces, Vernon thrusting his penis against Kevin's. The two males calmed down at once and began collecting the raisins that had been scattered around. Instead of screaming, they now made excited food sounds. All of us were relieved that it had been so easy. The keepers, always proud of their animals, claimed that bonobos are simply too smart to fight. An hour later the others were added without any problems. The new group consisted of the adult male, two adult females, two adolescent males, and the female infant. I have already described a few scenes from this enlarged group, such as the sexual interactions of the two females. In the course of a few weeks Vernon built up a close relationship with Louise. At first baby Lenore was jealous, jumping on the male's head and punching him in the eyes if he groomed her mother. Vernon might whisk her away, like a bothersome insect, but in general he behaved tolerantly. He grew more and more intolerant toward the young males, however.

After months of mounting tension it was decided to remove

the adolescent males. This was better anyway, as they are full brothers of the two breeding females. Normally, sexual interest between siblings is low, but not in the case of Kevin and Kalind. They had joined their sisters too late in life (after a period in the zoo nursery) to develop the natural inhibitions preventing inbreeding. The two males no doubt will be sent to other zoos to reproduce with unrelated females. In the meantime they are kept with the juvenile group.

One of the most interesting new relationships after the merging of two subgroups was that between Loretta (*right*) and Louise (*left*). The two females never fought, but Louise's daughter sometimes provoked tensions by grabbing at Loretta's food (which Loretta could hardly refuse without incurring Louise's anger). In this instance Loretta grins and gives the infant a hand, while moving the bundle of branches out of her reach. This grin is not, as in macaques, a sign of submission. Loretta clearly dominates the infant, but tries to appease her. In bonobos, the function of this signal comes close to that of a human smile. (San Diego Zoo)

Several months after formation of the larger group, Vernon became so hostile toward the two adolescent males that they had to be separated. Here Vernon (*left*) and Kalind in happier times. The adult male tickles the youngster in his armpit, making him choke with laughter. (San Diego Zoo)

When I revisited San Diego in the summer of 1985, the bonobos had a fine renovated enclosure, at least four times the size of the previous one. I was impressed by the improvements made by the animal-care staff. In the morning, before letting the bonobos out, the keepers would hide little pieces of food in the grass, under the bushes, and even under the ground. Raisins would be scattered about, and sunflower seeds would be poured into specially drilled holes in logs. Honey would be placed in an artificial termite mound, from which it could be extracted by sticking long grass stems or thin branches in the openings.

Environmental enrichment, as it is called, is increasingly popular at zoos. It prevents boredom and apathy in the animals, which is good for both their mental and their physical health. It also provides more interest for the public. The keepers feel that the enrichment program at the San Diego Zoo makes the daily rhythm of the bonobos more natural. Every morning they stay busy for hours digging in the sand, shaking the bushes, and working the honey mound.

It will be interesting to repeat my study on these "egg-hunting" apes. I watched them for several mornings to see how the discovery of food items stimulated erotic activity. Once Lisa, a new young arrival from Atlanta, hit on an orange while digging in the sand. Before uncovering the fruit, she looked around to check on what the others were up to. Kevin, now boss of the juvenile group, was approaching, unaware of her discovery. She immediately left the site to meet him, threw herself on her back, and grabbed him by the shoulders. During the ensuing mating Lisa squealed. Then she hurried back to dig up her prize. I do not know what Kevin would have done without this contact, but now he let Lisa keep her orange and even defended her when Kalind tried to snatch it.

It is hard to imagine bonobo society without sociosexual behavior; it perhaps would be like an engine without lubricant. Sexual conflict resolution is the key to bonobo social organization, and individuals learn its strategic value at an early age. It may be particularly important for females, thereby explaining their central social position, as reported by field-workers. A close, balanced relationship is created between the sexes, as well as among the females themselves (who have the unique GG-rubbing pattern unknown in chimpanzees). It may be that our human ancestors went through a similar stage of social organization before limiting the calming, cementing role of sex to family life. If a couple argue bitterly, then go to bed to seal the making-up process, they are acting very much as bonobos would.

Instead of inquiring which species, the bonobo or the chimpanzee, most resembles us humans, we can more fruitfully ask which elements of our social life are shared with the one or the other and which elements are uniquely ours. In nature unity and diversity go hand in hand, and our habit of exaggerating human uniqueness has led us to ignore the larger picture. We need to identify both the themes and the variations of the gigantic fugue of which we are a part. The themes are the traits we share with the apes, presumably owing to our common ancestor. The variations are the unique elements, which either evolved over the past few millions of years or were added by culture. The study of bonobos, while confirming the basic themes, has revealed a number of unexpected new variations, which inevitably will have a major impact on all future theories of human evolution.

CHAPTER SIX

Humans

Who can one hit, if not one's friends?
>—Sir Ralph Richardson to Sir Alec Guinness,
>   before punching him in the jaw

Studies of human aggression are to an unfortunate degree beset by another difficulty: investigators are urged, either by their sponsors or by their consciences, to find out what to *do* about some problem before they have formed any very clear opinion of what the problem *is*.
>—Paul Bohannan

If four different primate species routinely make up after they have fought, similar behavior in another, closely related species probably has the same origin. No one would object to this extrapolation if the fifth species were another animal, but because the fifth species I refer to is the "Crown of Creation," controversy is bound to arise. Animals are considered slaves of their instincts, whereas humans are believed to be creatures of intellect. The distinction is not so clear-cut, however. Animals do not respond automatically, and people are by no means free of deep-seated desires and emotions.

To understand the human species is a particularly challenging task. Because a truly objective view of one's own kind is impossible to achieve, it is not surprising that so many schools of thought and so many conflicting theories exist. Even though there is room for all these viewpoints, one approach meets with general hostility from scientists specialized in human behavior—the approach of the biologist. Yet it seems to me that if the biological perspective is so much at odds with all the others, there is all the more reason to consider it. It is not through ignoring divergent views that scientific progress is achieved.

The purpose of this final chapter is twofold: first, to stress the amazing lack of data on peacemaking in private human relationships; second, to reflect on this aspect of ourselves in a new and potentially enlightening way—by juxtaposing animal and human behavior. Because of my high regard for the psychological complexity of monkeys and apes, I believe that similarities to human behavior can arise in many ways, most of which we can only guess at. Two species may act the same because they share a long genetic history, because they have learned similar solutions to similar problems, for both reasons or neither. Hence, the parallels drawn here are not to be taken as proof that our actions are inescapable dictates of nature. Social behavior is in all species a blend of inborn tendencies, experience, and intelligent decision making.

Human behavior is unquestionably affected by the sociocultural environment. Golda Meir, former prime minister of Israel, once said in an interview with Oriana Fallaci that Palestinian schoolbooks pose arithmetic problems like this: "You have five Israelis. You kill three of them. How many Israelis are left to be killed?" We can hardly expect a desire for peace to develop in children fed on such hatred, and obviously this is exactly what is intended. On the other hand, the fact that parents and teachers can mold a child's attitudes is not a valid argument against genetic influences on behavior. One influence does not exclude the other. Many aspects of human behavior are so universal that they are best regarded as the combined product of biological raw materials and cultural modification, rather than as the independent invention of each culture. It is these raw materials, not the finished end products, that appear to be the same for the five primate species treated in this book, the special complexity of one species notwithstanding.

The Paucity of Knowledge

Three boys were interrogated at an Amsterdam police station after having drawn suspicion on themselves by spending more money than is normal for ten-year-olds. The youths admitted that they had found a wallet containing five thousand-guilder notes, but they had in their possession only a little over two thousand guilders. Where had the rest of the money gone? The answer made headlines. The boys had thrown two of the five bills into one of the city's age-old canals—their solution to the indivisibility of five by three. This is a dramatic illustration of how much people value good relationships.

I should qualify this. We value good relationships only to a point. The three boys must have been close friends. If one of them had been an outsider—new to the neighborhood, for example—an entirely different division might have been arranged. Who cares about an outsider, unless he is a very tough fellow? The goal of conflict settlement is not peace per se; it is the maintenance of relationships of proven value. This value is a highly variable item, not only across relationships but also across time within a particular relationship. Thus, a married couple who have successfully reconciled thousands of conflicts may nevertheless reach a point where it does not seem worthwhile to go through the same ritual again. They will increasingly place self-interest above marital harmony.

One compelling goal of people is to have relationships that work to their own advantage. If this occurs in perfect harmony, fine. If it requires coercion and threats, followed by soothing remarks, often this is fine too. Even if one party exerts constant pressure, we stay in the relationship as long as we need it. We do everything possible to keep our social network operative, not necessarily with the most agreeable methods. Some of the best relationships are riddled with squabbles, in that the two parties fluctuate between reinforcing their bond and getting the best possible deal from it. It is comparable to the way a drawbridge serves two kinds of traffic. Keeping the bridge down causes a jam-up of boats in the canal; keeping the bridge open brings auto traffic to a halt. Just as a drawbridge can never stay in one position, relationships continually go through ups and downs to ensure that issues do not remain unresolved and that hurt feelings are mended.

While aggression is part of each and every human relationship, social scientists treat it as an inherently evil behavior. "Aggression is arguably the most serious of human problems" is a typical opening sentence of books on the topic (this particular one is by Jeffrey Goldstein). Authors support such a statement by a review of derailed aggression and all the misery it causes. I am certainly not of the opinion that aggression is unqualifiedly good—I have seen my share of blood and injuries—but I would prefer that scientists take a more encompassing look. Beyond the excesses of murder, rape, and child abuse, there is an entire spectrum, including the everyday hostilities with which we are in fact quite comfortable. Rather than start with the assumption that aggression shapes our lives in a negative manner only, we would be wiser to leave all options open, including the possibility of constructive outcomes of conflict.

I speak from years of frustration with the literature on human behavior. How do people actually behave? Available are answers to questionnaires, which at best reveal how people perceive themselves and at worst how they wish to be perceived Available, too, are data on the behavior of human subjects in experiments. People who do not know one another are brought together in a laboratory room. All variables supposedly are under tight control in such settings, but the link with real life is lost. The observed social relationships have neither past nor future. We might as well investigate the swimming of fish by taking them out of the water. Where are the basic observations of human conduct within the family, at work, at school, at parties, on the street, and so on? Granted, there are methodological problems, but it should not be too difficult to take notes on people in action—not more difficult, surely, than fieldwork on dolphins or arboreal primates. In the natural sciences, simple descriptive data form the bedrock on which theories are built. Linnaeus preceded Darwin. The social sciences, however, seem to be trying to skip this tedious phase. Studies matching the descriptive detail of ethological work on animals are not easily come by.

Reconciliation behavior in humans is a case in point. Except for reports on preschool children and an occasional anthropological account, I am unaware of data in this area. It simply is not recognized as important. The subject indexes of major textbooks give abundant citations of "violence" and "aggression," but I have yet to find a single reference to interpersonal "peacemaking" or "forgiveness" (the clinical literature, which treats the process as it is mediated by therapists, is an exception). If the massive, well-funded aggression research of the 1960s and 1970s has failed to illuminate mechanisms of conflict resolution, it is largely because of a strong bias against the notion that aggression can be, even should be, *integrated* into our lives. During the era of Flower Power human aggression was regarded as a purely cultural product—and a highly undesirable one—the existence of which was entirely in our hands. To get rid of it, people needed only to control their material possessiveness, their desire for dominance, and their sexual jealousy. Why should humanity settle for the canalization, sublimation, or integration of such "diabolical" traits if eradication was within its power? Many social scientists were and are scarcely interested in checks and balances on aggression, simply because they refuse to believe that aggression is here to stay. Today in the 1980s, in the wake of total failure to shake off the unwanted heritage, we are still waiting for a revision of such optimistic theories.

I recently asked a world-renowned American psychologist, who specializes in human aggression, what he knew about reconciliation. Not only did he have no information on the subject, but he looked at me as if the word were new to him. I do, of course, speak with an accent, but this was not the problem. He reflected on my remarks, yet the concept had evidently never taken center stage in his thinking. His interest turned to irritation when I suggested that conflicts are inevitable among people and that aggression has such a long evolutionary history that it is logical to expect powerful coping mechanisms. He did not see what evolution had to do with it and argued that the most important goal is to understand and remove the causes of aggressive behavior.

To view aggression exclusively as an ugly, maladaptive trait requires that buffering mechanisms be ignored. If a mother monkey slaps her infant, then immediately embraces and consoles it, she has in one breath taught her child whatever she deemed necessary and demonstrated her continuing affection. The effect on the mother-child relationship is not necessarily what we think. For example, rhesus mothers, who are quite severe with their young, develop lifelong bonds with their daughters. Chimpanzee mothers, who hardly ever punish their offspring, rarely develop close-knit matrilines; most daughters migrate to other communities. If aggression were our sole criterion, we might call rhesus mothers "bad" and chimpanzee mothers "good." The judgment would be reversed if bonding were our favorite measure. And what if we preferred the loose bonds of chimpanzees over the close but strictly hierarchical ties of rhesus monkeys? The more we reflect on these issues, the less sense moral categories begin to make.

Am I, by trying to bypass the moral issues, condoning all forms of aggression? Do I believe that violent abuse is tolerable as long as it is followed by apologies, promises, or presents? Of course not. My point is that concern about the harmful effects of aggression is too narrow a basis for the study of such a broad behavioral complex. It is a matter of degree. We can handle some snowfall, not an avalanche. Until now scientists have looked at aggression as an avalanche. Anyone speaking of less disturbing, or even pleasurable encounters with it must, in their view, be deranged. I am convinced, all the same, that by opening our inquiries to include nondestructive forms of aggression, we may, in effect, also gain a better understanding of the forms that trouble us.

Our human societies are structured by the interplay between

Rhesus mothers treat their offspring quite strictly. This infant responds with a fear grin to his mother's bite; he had resisted her attempt to remove him from her belly. (Wisconsin Primate Center)

antagonism and attraction. Disappearance of the former is more than an unrealistic wish, it is a misguided one. No one would want to live in the sort of society that would result, as it would lack differentiation among individuals. A school of herrings is a good example of an aggregation predominantly based on attraction: the fish move together without any problems, but they have no social organization to speak of. If certain species, such as humans, reach a high degree of social differentiation, role division, and cooperation, this occurs because the cohesive tendency is counteracted by internal conflict. Individuals delineate their social positions in competition with others. We cannot have it both ways: a world in which each individual attains his or her own identity, and a world without clashing individual interests.

"When the focus of research is exclusively upon aggression, without measures of affiliation, there is a tendency to exaggerate its antisocial consequences," concluded Heidi Swanson and Richard Schuster from their experimental demonstration that cooperation is promoted, rather than hindered, by a moderate level of aggression among rats. Such research should not be limited to animals. It is time that we learn how people use aggressive behavior to reach their goals, and how they subsequently deal with the consequences. Insight into these processes will undoubtedly blur the distinction between positive and negative acts, for all acts are fused in the relationship and it is only the end result that counts. For example, I would not be surprised if reconciliations do more than merely rescue human relationships from undermining conflicts and tensions. Is not willingness to overcome hostile feelings the ultimate proof of commitment? Screaming and shouting followed by tenderness may actually strengthen a bond, in that the sequence assures both parties of the viability of the relationship. We do not trust a ship before it has weathered a storm. In the same way, a history of happy making up may give people the courage to be truly open with each other.

What makes the issue of forgiveness and reconciliation so intriguing is the paradoxes: bickering but cooperative rats, competitors unified in hierarchies, food struggles resolved through sex, battered wives attached to their husbands, the sympathy of hostages for their captors, and so forth. One explanation of the last riddle, given by Charles Bahn, is the emergence of feelings of extreme gratitude to someone who has made a credible threat on one's life without acting on it. In other words, terrorists who kill are murderers; those who almost kill are cavaliers fighting for a just cause, at least in the eyes of some of their victims.

Paradoxes disturb the neat dichotomies that we set up to clarify our thinking. For this reason, paradoxes are often treated as oddities. Still, their number may reach a size such that the dichotomization loses its usefulness. Evidently I believe that this has happened to the division between antagonistic and affectionate behavior. Not because of a lack of distinction— everyone can distinguish a slap in the face from a kiss on the cheek—but because of the intertwinement of the two in the long run. The condemnation of aggression as antisocial behavior is, like all morality, a simplification. If scientists do not detach themselves from such value judgments, they will never reach a full understanding of the way conflict shapes our social life.

Degrees of Sophistication

Monkeys and apes adapt their behavior to circumstances, achieving great sophistication in conflict resolution. They may not hold preliminary negotiations about the shape of the table at which the parties will meet, or set up so-called proximity talks with a go-between for delegations in different rooms, yet chimpanzees know what mediation is. In the Arnhem colony it is not uncommon for a female to break the ice between adult males who, after a fight, stay close to each other but seem unable to reopen communication. Avoiding eye contact, the two males play the familiar game of glancing over when the other looks away. A female may approach one male, briefly groom or touch him, and walk over to the other with the first male following closely. This way he need not face his adversary. When the female sits down next to the second male, both groom her. Only a small shift is necessary for them to groom each other after the female has walked off. That the mediator knows what she is doing is clear from the way she looks over her shoulder and waits for a male who is reluctant to follow. She may even go back and tug at his arm.

Although I have never observed conflict mediation among macaques, this is not necessarily because of a lack of social awareness in these monkeys. Once the second-ranking rhesus male, Hulk, chased one of the younger males, Tom. Immediately afterward, Tom's mother approached Hulk to groom him. While she was doing so, Tom came closer and closer until he sat less than a meter behind the two. As soon as his mother noticed him, she stepped aside and looked away. She left the scene when her son took her place against Hulk's back. We have witnessed a number of similar situations, in which monkeys made room for contact between former opponents. These observations warn that the mediation skills of chimpanzees and humans may not be totally without antecedent. Our monkeylike ancestors may already have possessed an important prerequisite—the ability to recognize and facilitate reconciliation attempts between others.

Loss of face is a calamity that we humans easily recognize, yet find hard to define in objective behavioral terms. I am convinced that face-saving tactics are as important among our simian relatives as they are among ourselves. If two male chimpanzees are reluctant to make up, but without hesitation grab the opportunity to make an approach behind a mediator's back, it seems as though pride has prevented initiatives of their own. Occasionally males solve this problem without help from a third individual. Yeroen, for example, would feign interest in a small object to break the tension and attract his adversary. He would suddenly discover something in the grass and hoot loudly, looking in all directions. A number of chimpanzees, including his adversary, would rush to the spot. Soon the others would lose interest and leave, while the two male rivals would stay. They would make excited sounds as they sniffed and handled the discovery, focusing all their attention on it. While doing so, their heads and shoulders would touch. After a few minutes the two would calm down and start grooming each other. The object, which I was never able to identify, would be forgotten.

The principle of a collective lie is that one party deceives and the other acts *as if* deceived. It is tempting to interpret the foregoing incidents in this manner. The fact that, in addition to Yeroen himself, his rival was fascinated by a discovery that induced so little interest in the others suggests that both males understood the purpose of their actions. In humans, collective lies are a familiar face-saver. Colin Turnbull described a beautiful example in the BaMbuti pygmies of the Congo. Among these forest people it is always the women who build the huts, so they are able to make a point during matrimonial disputes by demolishing part of their home. Usually, the husband gives in when a fight escalates to this level. One time, however, a particularly stubborn man did not stop his wife and even remarked to the camp at large that she was going to be dreadfully cold that night. To avoid being shamed, the woman had to continue the destruction. Slowly she started pulling out the sticks that formed the framework of the hut. She was in tears because, according to the anthropologist, the next step would be for her to pack her belongings and return to her parents. The man looked equally miserable. Things were clearly getting out of hand, and to make matters worse, the entire camp had come out to watch. Then the man suddenly brightened and told his wife that she could leave the sticks alone; it was only the leaves on the roof that were dirty. She gave him a puzzled look, then understood. Together they carried the leaves to the stream and washed them. Both were in a much better mood when the woman put the leaves back on the hut, and the man went off to hunt food for dinner. Turnbull comments that although no one believed the pretense that the woman had been removing leaves because they were dirty, everyone played along. "For several days women talked politely about the insects in the leaves of their huts, and took a few leaves down to the stream to wash, as if this was a perfectly normal procedure. I have never seen it done before or since."

Collective lies allow compromises to be reached without creating definitive winners and losers. It is the opposite strategy of an explicit reconciliation, during which both parties openly refer to the matter that divides them. Excuses for rapprochement add an extra layer of intentions to the peace process. When we peel back the layer of declared motives, we may find a very different set of motives. In humans, the hidden motives are usually less noble than the ones presented to the outside world; self-interest is at the root of virtually every olive branch. What we discover may even be positively malicious. Individuals may go so far as to feign a conciliatory mood in order to reach exactly the opposite objective: revenge. Among the Arnhem chimpanzees this extreme form of deceit occurred on six separate occasions in the years that I watched them, all perpetrated by adult females who had been unsuccessful at catching their opponent during a previous aggressive incident. The female would approach her escaped victim with an invitational gesture, such as an outstretched open hand, and maintain her friendly attitude until the other, who was attracted by it, had come within arm's reach. Then the female would suddenly grab and attack her naive opponent.

Instead of calling this a deception, we could cite the alternative explanation that the female had changed her mind; that she really had wanted to make up, but that when her opponent came close, hostile feelings flared up again. This interpretation has weak points, however. Why were the victims in all cases low-ranking individuals capable of outrunning the female? Why did she wait until the last split second to change her mind? And why did she need to physically punish her victim, when a soft grunt would have sufficed to stop the approach? My impression is that the attacks were much too abrupt and vicious to have resulted from hesitation and conflicting emotions. I believe, in short, that these were premeditated moves to square an account. The chimpanzee's capacity for pretense is supported by other observations, both in captivity and in the wild, and by experimental research.

The above anecdotes make clear that a connection between human and animal behavior by no means implies that our conflict-resolution skills are "instinctive" in the narrow, colloquial sense of the word, that is, inborn stereotypical behaviors that we perform without thinking. If our fellow primates apply so much intelligence to these situations, would humans not do the same or more? Foresight and planning permeate all segments of our social life, including the way we deal with tensions and aggression. I still remember how, as a child, I would hurry to make up with my younger brother if I heard one of my parents coming, realizing full well whom they would side with. And my older brothers did the same when they had squabbled with me. Since early experiences never leave us, I immediately recognized the mechanism when I saw it in a chimpanzee family at the Yerkes Primate Center in Atlanta, where I recently conducted a study.

The family there comprised a female named Lolita and her two offspring: a fully adult daughter, Sheila, and a six-year-old son, Brian. The three of them lived in a group of twenty chimpanzees. Although Lolita is, by chimpanzee standards, fairly petite, she is the colony's alpha female (perhaps because she is the oldest individual). Unlike her mother, Sheila is unpopular in the group. She proved extremely selfish during the tests that I arranged to record food-sharing behavior, and she was the favorite target of two boisterous adolescent males when they were in the mood to test their fighting skills on females. One of these males was her younger brother, Brian. When his pal was around to back him up, Brian frequently teased Sheila by throwing sand, spitting at her, or giving her an unexpected poke in the back. Obviously, this did not go over well with Sheila. If she encountered Brian alone, she would push him if he slept, refuse to groom him if he invited her, or act in other subtly negative ways that sometimes led to a quarrel. Although Sheila still physically dominated her brother, she had to be careful. As soon as Brian gave a little scream, Lolita would look up. I never saw her in a hurry to correct the situation, but she would keep an eye on her children and often approached the scene. She would sit down a few meters away, diplomatically acting as if nothing was the matter. This was just the kind of pressure to make up that Sheila needed. She would embrace Brian, groom him, or pull at his leg with a play face (she never usually played). All the while the two would throw glances at their mother. Only twice did Lolita actually interfere. Both times Brian took advantage, helping his mother chase his big sister.

Strategic reconciliations are quite common among chimpanzees. In Arnhem, Nikkie would make peace in the midst of a conflict with his coalition partner if the third male started an intimidation display. At the Yerkes Field Station, I observed a few unusually quick reconciliations between females who had had a fight before the keeper arrived with the bundle of branches that I used for my food-sharing tests. Upon seeing the keeper, the female rivals hurriedly kissed and embraced. I suppose neither wanted to run the risk of not being on friendly terms in case her rival got the food.

In short, several basic variations on the theme of peacemaking—including third-party mediation, opportunism, and deception—can be found in both humans and chimpanzees. No doubt humans surpass apes in their degree of sophistication, taking into account more options and consequences when deciding whether to settle a dispute. The salient point, though, is that both species make decisions based on experience and calculation. For this reason the observed similarities may have more to do with the way the brain solves problems than with the genetic programming of behavior.

Conflict resolution in monkeys seems a simpler and more straightforward process. But in comparing them with humans and apes, we should not stress contrasts at the expense of continuity. All five primate species seek contact with former adversaries. They do so in entirely different ways, ranging from the GG-rubbing of female bonobos to culture-specific human patterns, such as an aloof handshake. Each species applies all the social awareness and intelligence at its disposal. The complexity of the approach can range from a simple grooming contact between two rhesus monkeys to the typically human strategy of testing, via intermediaries, the feelings in the opposing camp before representatives of the two sides meet.

Only a few ingredients of reconciliation need to be inborn for the mechanism to work. An absolute minimum requirement is, of course, individual recognition; members of the species have to be able to remember with whom they fought. Other necessary ingredients are the ability to make fairly rapid emotional shifts from anger to friendliness, and the ability to be soothed by body contact and certain gestures, such as withdrawal of the lips from the teeth in a grin or smile. But even these aspects are influenced by the environment. For example, a monkey raised in isolation will be thoroughly disturbed the first time he or she is touched. So the search for the "immutable bedrock" of reconciliation is a bit like the search for the Holy Grail. It is much more profitable to think in terms of *potential*. With our simian relatives we share a psychological template that, filled in through interaction with parents, siblings, and peers, allows us to develop the social skill of reconciliation.

The possession of this template is not self-evident, and nature has produced it in different shapes depending on the environment and life-style of the species. The characteristics of the human template are undoubtedly related to our long history as hunter-gatherers. In view of the close-knit community life and strong interdependency among extant hunter-gatherers, we can speculate that the capacity to find alternatives to overt aggression, and to restore the social fabric after disruption, must have been of critical value in human evolution.

Conditions of Peace

"The most general effect of fighting is to cause an injured animal to move away from another, with a resulting regulation of the use of space," wrote John Paul Scott, voicing the traditional idea that aggression leads to dispersal. My observations attest that this rule does not hold for group-living primates. Nor do these primates depend on time to "heal their wounds." They have means to speed up the process. From fellow primatologists I have heard of three recent, as yet unpublished studies confirming this for other species. Now that the phenomenon has been demonstrated, we need to investigate the *conditions* that determine whether individuals reunite after a fight, enter a spiral of escalation, or ignore the damage to their relationship. The choice undoubtedly depends on a multitude of considerations, such as the value of the relationship, its history, and the cost of holding a grudge.

Peace is always linked to particular conditions. In international politics nation A is willing to be friends with nation B only if B stops supporting certain rebels, withdraws its troops from a third country, compensates A for war crimes, returns this or that to A, accepts A's claims to a border region, is prepared to help A against its enemies, and so on. The stronger a nation's position, the higher its demands, because the preferred settlement is one that lets A decide B's future behavior. In addition to bringing the end of war, a peace treaty marks the beginning of the relationship on new terms. It is these terms that are in every leader's mind—and not just after the war has come to an end. Before the Italian army decided to join Hitler's invasion of France in 1940, Benito Mussolini reportedly told his marshals, "I only need a few thousand dead so that I can sit at the peace conference as a man who has fought."

The motivation for peacemaking among primates is still largely unexplored. Rhesus monkeys mostly reconcile with relatives and members of their own social class. Since these are usually also their supporters, it is not hard to guess the reasons for making up. Chimpanzees, on the other hand, exhibit a tremendous sex difference, with males being both more aggressive and more conciliatory than females (which I have explained as related to the flexible coalition network of males and their need to remain unified in the face of intercommunity violence).

The mechanism allowing male chimpanzees to achieve social

integration in spite of fierce competition is their formalized rank order, which facilitates peacemaking through its role division. Hence the best predictor of improvement of an adverse relationship is the emergence of an unequivocal victor. I once invested several hundred hours in recording an ongoing dominance struggle between Yeroen and Luit in the winter quarters of the Arnhem chimpanzee colony. My objective was to be there when one of the two males would formally submit to the other by means of the familiar bowing and pant-grunting ritual, and to compare the relationship before and after this moment. After three months of daily intimidation and noisy conflict, Yeroen finally capitulated. I do believe that I witnessed his very first submissive grunts to Luit, because the other chimpanzees responded by rushing to the two males to embrace them. The group must have been waiting as eagerly as I had. My data showed a sudden, dramatic improvement of the relationship between the two rivals. In the following week they groomed each other twenty times more than during the period preceding Yeroen's acknowledgment of Luit's status, and their confrontations rapidly decreased in frequency and intensity.

Does acceptance of inequality alleviate tensions between humans also? Again, this is an area of human behavior in which, to my knowledge, no data are available. The fabled rivalry between Anatoly Karpov and Gary Kasparov, two Russian masters contending for the world chess title, parallels the Yeroen-Luit struggle. Chess players need tremendous concentration, so they sometimes (as in this instance) perceive their opponent's body movements or clothing as deliberate disturbances. The hostility, suspense, and accusations between Karpov and Kasparov mounted with every match, and there were many. It took an unprecedented ninety-six games before a decisive outcome was reached. After the final move on the board, which sealed the dethronement of world champion Karpov, the two men rose from the table, shook hands, and chatted briefly. Nothing unusual? It was the first observed friendly contact in two years. The battle decided, they could at last afford some warmth.

Equality and unity are hard to combine within one social

system. In the absence of a hierarchical organization, internal strife leads to fissions. Leftist political movements that have tried for ideological reasons to organize themselves without elected leaders have tended to break into splinter groups. And compare the Roman Catholic Church with Protestant religious groups. In spite of a fair amount of internal discord, Catholics are unified under a central authority, whereas Protestants organize themselves essentially in territorial fashion, with a multitude of dissenting sects. In view of this contrast one wonders if the ecumenical ideal, reconciliation among all Christians, can ever be attained without adoption of a pyramidal structure. Between the Anglican Church and Rome such a process seems to be under way, and it is interesting to follow the preparatory moves. Despite the obvious fear of absorption by the so-called mother church, the Anglican archbishop Runcie has pointed to the usefulness of the papacy as "a focus for unity and affection" (*Time*, June 7, 1982). If a deal is in the making, its hierarchical outline is already recognizable.

This is not to say that egalitarian conflict resolution is an impossibility. It has been described for pair-bonded gulls, for example. According to Judith Hand, gull pairs solve conflicts over food by sharing large items and consuming small items on a first-come-first-serve basis. It is a simple convention, as simple as the priority convention of a rank relationship. One can see the resulting emergence of what Hand calls spheres of dominance within a relationship. In this more flexible form of conflict resolution each partner at times yields to the other, depending on the issue at hand. Many a married couple follows this scheme.

Female bonobos present us with another possible exception. From what I have witnessed in the San Diego Zoo and gathered from field reports—both rather preliminary sources—the females of this species get along remarkably well despite the absence of a pronounced hierarchy. Do their intense sexual contacts serve as an alternative? How effective is this sort of social organization compared to a hierarchical one? The initial advocacy of lesbianism by the feminist movement comes to mind, as it was intended to create solidarity. Is it working out that way?

We need to study such alternatives to know the whole range of unifying social mechanisms, remembering all the while that the oldest and most widespread method, adopted by both males and females of many species, is hierarchy formation.

Unification through subordination has shaped the world. Even though wars temporarily divide people, historically they have been a unifying force. Most modern nations owe their existence to a few lesser conquerors and one great one. An exception to this rule is several former colonies, such as the United States and India, which achieved their sense of unity mainly during a period of insubordination, while struggling for freedom. This sort of cooperation tends to erode after the colonial power has left the scene. Thus, it is uncertain whether the Republic of India will manage to cohere (Pakistan and Bangladesh already have split off), and America's land mass would today probably accommodate two independent nations had the North not defeated the South in a civil war.

If in human military history submission is often followed by integration, there is a mirror image, namely that wars between groups who once were united can be especially vicious. Napoleon Chagnon studied the Yanomamö Indians of Venezuela, whose males wage bloody wars over crops and women. The anthropologist observed, "A village's mortal enemy is the group from which it has recently split." This may also be true at the interpersonal level—after a split in a large family, for instance, or during a divorce. We have to be careful, though, because our perception could be biased. Perhaps hatred *looks* worse to us when it involves previously bonded individuals than when it involves strangers. The brutal violence between the two chimpanzee communities in Gombe particularly shocked the field-workers, because the aggressors and their victims had in earlier years lived as close and friendly members of a single community. "By separating themselves," says Jane Goodall, "it is as though they forfeited their 'right' to be treated as group members—instead they were treated as strangers." My question is, *Just* as strangers, or worse than that because of previous familiarity?

I am reminded of an incident that I documented in detail as a graduate student in 1975. The alpha male of a captive group of long-tailed macaques showed a deep gash, caused by another male's canines. He literally trembled on his legs each time his adult son, who had grown to about twice his own size, walked by. Obviously there had been a fight between the two, yet the younger male never claimed the top position. It almost seemed as if he did not realize that he could; perhaps he had bitten the older male in self-defense, not as a challenge. At any rate, the alpha male's nervousness created great tension, which the two males in typical macaque fashion took out on a scapegoat. Together they would chase the scapegoat into a corner, then take turns in the assault. After several days they calmed down, the alpha male maintained his position, and I could write my very first article, which dealt with the stabilizing effect of a mutual target for aggression. The only aspect I could not explain was that the scapegoat was not one of the lowest-ranking monkeys, which is the usual choice; it was the mother of the younger male, and he himself initiated most of the attacks on her.

After quite a bit more experience with primates, I now believe that this was not an exceptional choice at all. Redirection of aggression often occurs at the expense of close kin and friends. As humans, we all know the phenomenon of one spouse who takes the strains and frustrations of the job out on the other spouse. As long as this can be absorbed within the relationship and does not turn into violence, it is a fairly safe procedure safer, perhaps, than venting such feelings at the workplace. The result is yet another paradox, however; namely, that anger is most readily expressed against persons with whom we know we can make up. We are taking advantage of the fact that they love us.

There is another side to this issue. If people are insecure about their relationship, anger may go unexpressed in order to preserve a fragile peace. In very general terms, then, whether or not we show aggression depends on the expected impact on the relationship, and whether or not we make peace depends on the demands of the other party. If aggression goes too far, we endanger a marriage or friendship; if peace is offered too soon, we end up on disadvantageous terms. How these two tendencies are balanced is little understood, but since the same ambivalence is recognizable in other primates, detailed studies of conflict resolution in animals may well lead to theories that can be tested on human behavior.

Children

Children irresistibly remind us of monkeys when they climb trees, race around the house, or play rough-and-tumble. There is nothing wrong with this comparison as long as it concerns play behavior and motor skills. In academic circles, however, a peculiar inversion of the child-primate comparison has arisen: the notion that other primates are *mentally* like human children. The idea that all primates except adult human beings belong in the kindergarten is just too convenient to be true.

The main source of confusion is films. Hollywood loves simian actors. Moviegoers, who see them as grotesque imitations of human beings, cannot get enough of their grimaces and antics. To me, *Bonzo* movies and calendars with pictures of dressed-up apes are anathema. They are an insult to the intrinsic dignity of these creatures, and I do not blame the "actors" for occasionally attacking the crew with which they work, for their training is not free from physical punishment.

Decades ago John Bauman used a dynamometer to compare the muscle strength of adult chimpanzees with that of football players at the local college. The young men had an average one-handed pulling strength of 79 kilograms and a maximum of 95 kilograms, whereas the apes easily pulled several times that weight. One male chimpanzee who weighed 75 kilograms achieved a one-handed pull of 384 kilograms. As a result of their intimidating strength, adult chimpanzees are not seen in direct interaction with people in movies or on television. Apes in the entertainment business are rarely older than seven or eight years, which is comparable to a human child of ten. For the same reason most language experiments, in which apes are taught to communicate by means of symbols or hand signs, are terminated when the subjects reach puberty. It is not surprising, therefore, that in the public mind apes never grow up. They are seen as cute and playful creatures who let themselves be carried around and constantly are into mischief. Even scientists are misled, as evidenced by the frequent claims that the mental development of apes does not surpass that of a human child of x years of age. The value of x varies but usually does not exceed six. G. Ettlinger has proposed that "certain theoretical issues may be resolved if the comparison between primate and man is re-specified: primate/human *child*" (italics in original). It is remarkable, though, how rarely this opinion is heard from scientists who have had first-hand experience with adult apes.

To categorize a particular group of adults as childlike is not new, and I am suspicious of the motives. White men have shown this paternalistic attitude toward other races, toward women, even toward entire countries. (Hear General William Westmoreland: "Vietnam reminds me of the development of a child.") According to Stephen Jay Gould, the "primitive-as-child" argument helped to justify slavery: "For anyone who wishes to affirm the innate inequality of races, few biological arguments can have more appeal than recapitulation, with its insistence that children of higher races (invariably one's own) are passing through and beyond the permanent conditions of adults in lower races." Application of the same argument to our anthropoid relatives is inappropriate for two reasons.

In the first place, adult apes are not known for their cooperation with humans, so our current knowledge of ape intelligence and psychology is almost entirely based on experiments with youngsters. These individuals are about as representative of their species as preschool children are of the human race. Adult psychology is vastly different in both species. It revolves around status, sex, means of livelihood, and progeny. The calculated power games of adult male chimpanzees and the mediation skills of adult females, to mention just two examples, reflect preoccupations and social awareness much more comparable to those of men and women than to those of children. The collaborative castration and elimination of a rival for reasons of sexual competition, as witnessed in the Arnhem colony, is in every respect an affair of grown-ups.

From human families who have raised apes we know that the young of the two species make excellent mutual playmates. They play the same sort of games (king of the hill, blindman's buff, tickling matches), and show the same carefree attitude. Their sense of fun, communication, and even taste in television programs match perfectly. The children do not automatically outsmart the apes. When Winthrop and Luella Kellogg took hundreds of standard measures on the growth and development of their son, Donald, and a female chimpanzee, Gua, they found the ape doing better than the boy. She ate with a spoon, drank from a glass, and announced her bladder needs (by slapping her genitals with her hands) at an earlier age. Even in word comprehension and many of the intelligence tests, Gua was ahead of Donald. When both youngsters were about eighteen months old, the observations were terminated. One might argue that major differences, in favor of the child, would have arisen had the experiment lasted longer. To this the Kelloggs answer, "If we are entirely open-minded on the subject, we can hardly overlook the logical possibility that the ape might continue to demonstrate a superiority in many outstanding ways." Except for language ability, which is known to be better developed in our species, the Kelloggs' view, expressed in the early 1930s, still stands.

The second reason why we should think twice before reducing a complex comparison between two species to the simple apes-are-just-like-children conclusion is that the human species is considered to be *neotenous*. Compared to other primates, humans have a retarded maturation and retain some juvenile traits into adulthood. For example, our high forehead, large brain size, and sparsity of body hair are more typical of infant apes than of adults. Outside biology, the neoteny argument is not always understood; it has even been invoked to support precisely the claim against which it so forcibly argues.* What it really means is that human beings resemble, in the famous words of Louis Bolk, "a primate fetus that has become sexually mature." Others have extended this idea to include youthful behavioral characteristics, such as the remarkable playfulness and curiosity of *Homo ludens*, as Johan Huizinga called our species. In summary, it is probably closer to the truth to say that humans look and act like ape children than that apes look and act like human children.

Nevertheless, comparisons between humans and other primates involve, on the human side, mostly child behavior. An increasing number of researchers now apply ethological observation techniques to the study of children, whereas the field of adult human behavior is still largely untouched by this change in methodology. Certainly it is easier to watch children than adults, who tend to become self-conscious when stared at by someone taking notes each time they smile, raise their voice, hide their face, wipe their forehead, laugh, or slam the door. Children go about their business pretty much unaffected even if a whole team of ethologists follows them. Fred Strayer, who is in charge of a long-term investigation at a community day-care center in downtown Montreal, has a large staff that videotapes all sorts of activities, from meals to free play. Having started his career in primatology, he sees his present work as basically similar to monkey watching except for two differences. Children talk, so that an elaborate system to classify content and tone of verbalizations had to be designed. Second, it is hard to obtain direct information on one highly important segment of the children's lives, the time they spend at home.

Strayer is interested in the theories of Michael Chance, who postulates that the cohesiveness and internal coordination of social groups depend on the focal position of dominant members. Subordinates are attracted to, closely follow, and imitate

*For example, political scientist Glendon Schubert wrote in 1986, "Given the neoteny of contemporary humans, it should hardly be surprising that the behavior of our children has the most in common with the behavior of adult chimpanzees or baboons."

individuals at the top of the social ladder. In support of this model, known as the *attention structure,* the studies in Montreal indicate that hierarchies determine the selection of friends. Already at one year of age, conflicts among children have a predictable outcome. A dominance order is easily recognizable, although it is still subject to change. As the children grow older, their rank order stabilizes and begins to affect social attraction. By the age of four, dominant children have become the preferred playmates and friends. From then on, high rank among peers is associated with popularity. Because of the structuring effect of the rank order, and the way dominant children use their position to stop fights among others, Strayer has been one of the first to speak of the "prosocial" functions of aggressive behavior.

But it cannot be said that aggression never disturbs social relationships. Hubert Montagner, conducting similar research in France, has made a sharp distinction between two types of dominant children. One category he calls "aggressive dominants." These are mostly boys, who go around bullying their peers. They hit or push others without reason, claim their toys without asking, and otherwise disturb the harmony. The second category are the so-called leaders, among whom there are as many girls as boys. In contrast to aggressive dominants, leaders warn first—and wait for the other's response—before using force, which they rarely do. Another difference is that they make up after fights and otherwise employ calming gestures. For example, if new children arrive at the day-care center, leaders approach and comfort the little ones and threaten those who make them cry. Obviously it is these diplomatic dominants who enjoy popularity, not the bullies.

One particular peace gesture is unknown or very rare in other primates, yet common in our own species. It is the offering of presents.* In adults it takes the form of sending flowers, arrang-

---

*Outside the primate order gift giving is quite common. Food transfer is part of the courtship display of many birds, and some insect males bring a "nuptial gift" when approaching a female. These gestures serve to turn a potentially hostile situation between territorial or predatory animals into a cooperative one.

ing a conciliatory dinner, or (among the wealthy) buying jewelry. Irenäus Eibl-Eibesfeldt's studies in a wide range of human cultures indicate that gift giving and food sharing develop spontaneously, without training, in young children. Montagner emphasizes the importance of this behavior after disagreements. If one child brings a toy to another with whom he or she has just fought, the two typically engage in friendly contact, a joint effort, or a game of mutual imitation. Gifts serve to repair ties. If there are not any suitable presents around, no problem. Children are good simulators. They search in their pockets, as if they have lots of things with them, and offer an empty hand to the other—who happily looks at the imaginary gift.

Observations by Reinhard Schropp in a German kindergarten have shown that gift giving is most common between children who do *not* have close ties. It is used as a way of initiating contact. One advantage, according to Schropp, is that the attention is focused on the object. The children hardly need to look at each other, which lowers the risk of losing face if the object is rejected. If accepted, on the other hand, the object provides a convenient topic for further interaction.

Another interesting face-saving technique was observed by Harvey Ginsburg among American children on a playground. Fights were usually broken off if one party adopted a lowered body posture while avoiding eye contact. The defeated child would sit on hands and knees for a brief moment, sometimes tying his shoes. Ginsburg saw the shoe tying as an excuse, not as a problem of loose laces, because during play activities laces apparently stayed tied all the time. Furthermore, a boy once interrupted a fight to "tie" his loafer (a shoe without laces). The investigator speculated that shoe tying, while signaling submission to the opponent, also contains a message to the onlookers: "If only my shoes would stay tied, I really could win this fight!"

In 1984 Steve Sackin and Esther Thelen published a short technical paper entitled "An Ethological Study of Peaceful Associative Outcomes to Conflict in Preschool Children." This was the first "naturalistic" study comparable to my work on monkeys and apes. It concerned American children at two day-care centers, ranging in age from five to seven years. One hundred sixty-five conflicts in which teachers did not interfere were recorded. Basically, these encounters had two outcomes: either the fight ended with submission by one child, and the two opponents separated; or it ended with an exchange of friendly behavior, and the two opponents stayed together. Reconciliation took the following forms, given here in descending order of frequency, with the authors' definitions:

• *Cooperative propositions*—statements of friendly intent and suggestions for collaboration, such as "I'll be your friend" or "You can help me build this house."
• *Object offering*—already discussed.
• *Grooming*—handholding, stroking, kissing, embracing, or other forms of touching.
• *Apology*—verbally expressed regret at the result of the fight.
• *Symbolic offer*—a promise, such as "I'm going to bring you my truck."

Except for grooming (which occurred one out of five times), these patterns are uniquely human. It was also found that the frequency of reconciliation depended on the preschooler's sex. Fifty percent of the fights between boys ended with the opponents close together, as did 40 percent of the fights between girls, but only 12 percent of the fights involving a boy and a girl. Note that this is not the same sex difference as in two of the four primate species I have studied. Boys and girls reconciled more or less equally often, but almost exclusively with children of their own sex. This is not so surprising in view of the well-known preference of young children to form same-sex friendships.

Let me point out again that children are not the best possible standard against which to measure monkeys and apes. Many aspects of human social life change dramatically in adolescence, especially of course the relation between the sexes. Studies on our young are extremely important, however, if we wish to understand how peacemaking skills are acquired, and especially if we wish to guide this process. One of the questions we might ask is to what extent teachers should interfere in children's social affairs. The children that I host each year for a week-long introduction to monkey watching always tell me, as we discuss reconciliation behavior at length, that they hate to be pushed in the direction of their rival with the suggestion to shake hands. They do not believe in forced forgiveness, at least not outside the family.

To answer questions of this sort, I intend to compare juvenile stump-tailed and rhesus monkeys. Rhesus mothers are the interfering type, always on the alert when their offspring are in conflict with peers. Stump-tailed mothers take a more relaxed attitude: groups of juveniles are permitted to play, fight, and reconcile by themselves, unless the aggression gets really out of hand. Is this one of the reasons why adult stump-tailed monkeys deal so much better with social tensions? In human children, too, supervisors need to strike a balance between prevention of excessive violence and overregulation. The basic rules of peacemaking cannot be learned if fights are always broken up before a decisive outcome has been reached. In my recollection, such action only leads to additional rounds after school.

Even more important is the *example* set by adults—not how we tell children to behave, but how we ourselves behave after an outburst of anger. Children are first-rate observers who pick up the smallest change in facial expression. Should we allow them to witness entire sequences of marital dispute, from reproaches to apologies, or is it better to hide disagreements? Opinions are divided, and there are few data to guide parents, except of course that physical violence within a family thoroughly disturbs children. Mark Cummings and coworkers report upset reactions of young children to verbal fights as well, with some of them attempting to comfort or reunite the angry parents; but whether such disputes should be avoided totally in the presence of children is hard to decide.

The young chimpanzees of Arnhem had no lack of dramas to watch. They were even present during power struggles among adult males, which sometimes involved the entire colony. They would hang upside down under their mother's belly for as long as the pandemonium lasted, to be set free only after everyone had calmed down. Juveniles were remarkably attracted to reconciliations. They would watch from the sidelines during the tense preliminary moves, but immediately respond to the final embrace by excitedly jumping on both parties, or circling them while hooting.

One way to investigate the effect of the social environment on the development of conciliatory behavior might be to have a rhesus monkey grow up among the stump-tailed monkey group. The two species are close enough that an adoption could be arranged. It would provide the rhesus infant with completely different models for conflict resolution. Not that our subject would turn into the greatest peace advocate on earth— the feisty rhesus temperament must be partly inborn—but even moderate effects would provide a basis for further experimentation to determine the factors shaping conciliatory behavior and social tolerance. Educators could learn from such research; if monkeys are malleable in this respect, it is more than likely that human children are as well.

In addition to reunions between former adversaries, a second important mechanism is the reconciliation of conflicting interests *before* things get out of hand. Observations on food sharing among apes, and the role therein of reassurance behavior, indicate that preventive conflict resolution is not limited to our species. That I have relatively few data is largely because potential conflicts are not as easily defined as the postconflict situations on which my studies have focused. It would be fascinating to study how human children learn to anticipate conflict and negotiate simple deals to forestall problems ("You can play with my doll, if you give me some marshmallows"). Especially at this point, language makes a difference.

Cultures

"Man and culture originated simultaneously; this by definition," asserts Leslie White in *The Evolution of Culture*. Since he

estimates the origin of culture at one million years ago, it follows that our ancestors were, in his opinion, animals before that time. The key to culture, according to White, is our symbolic faculty. In a footnote he admits that "a portion of the behavior of man is not symboled and is therefore not human," but the only examples he provides are coughing, scratching, and yawning. In his extreme view, then, we are entirely our own creation: we are what we want to be.

For a biologist the idea of infinite cultural flexibility is unacceptable. When I visit a foreign culture, I am always struck by the familiarity of everything: the way people laugh, how they argue and about what, the way young men look at young women and vice versa, the change in a mother's voice when she talks to her baby, the strutting of important men, and so on. I am among my own kind. A cultural anthropologist making the same trip will focus instead on unique concepts in the language and on peculiar habits, clothing, and social institutions. He sees many striking differences and arrives at the opposite conclusion from mine: these people may yawn and cough like the rest of us, but that is where the similarities end.

Slowly, amid great reluctance, these two points of view are now moving closer to each other; there is obviously truth in both. Ethologists have discovered that many animals develop local behavioral traditions (for example, some bird species have different song "dialects" in different regions), and that revelation has sensitized us to human cultural variation. On the other hand, recent cross-cultural studies have demonstrated that certain aspects of human behavior are too universal to be entirely culture dependent (for example, in a large majority of societies boys behave more aggressively than girls). There still is a long

Two simultaneous reconciliations after a quarrel among juvenile stump-tailed monkeys. Two of them engage in a mount (*top*), and the other two in a hold-bottom. In this species adults leave youngsters pretty much alone to play, fight, and make up among themselves. (Wisconsin Primate Center)

way to go before biologists and cultural anthropologists will drink out of the same cup, but perhaps the generation of tomorrow will cling less tightly to the opposing dogmas of today.

Culture clearly affects human social life everywhere, including the way we control aggression and make peace. Yanomamö Indian women cultivate a magical plant, the leaves of which are thrown on the men when they have club fights. These ritualized battles provide an outlet for the high aggression level that Yanomamö men need to maintain in order to be taken seriously in their culture. The leaves, which allegedly keep male tempers under control, prevent club fights from escalating to shooting matches.

Americanized Hispanic women say they miss the nightly serenades customary in their culture. In Mexico it is quite acceptable for a man to wake up an entire street to declare his love, or to surprise his mother. Usually he does not perform himself; a professional singer with guitar, a trio, or an orchestra is hired for the job. Serenades also serve to ask forgiveness and repair a marriage. A more public way of making up is hard to imagine.

Villages on the island of Bali have a special hut to which people are sent to settle disagreements. The hut, located in a field outside the village, is composed simply of two poles with a roof on top. The absence of walls allows villagers to keep an eye on the two troublemakers. They sit with their backs against the poles, which are only a few meters apart, and may not return before their differences are healed.

The large family of Oscar Bimwenyi Kweshi celebrated its reunion in Zairian fashion after having been bitterly divided for years. The entire family met in a farmyard to listen to a public acknowledgment of errors by the brother and sister who had caused the split. After this, a chicken was sacrificed and both parties poured wine into a calabash, which was then shared as a reminder of the breast from which they had all fed.

Among the head-hunting Kiwai-Papuans of New Guinea, one village signals its desire to end a war by laying a branch across the path to the enemy village. If their offer is accepted, the men approach the village with their wives walking a few paces ahead. Bringing women along signifies good intentions. The reception is friendly, gifts are exchanged, and the men mutually break each other's beheading knives. At night the hosts sleep with the visiting women in order to "put out the fire," as it is known. The same happens on the return visit, and the war is declared over.

One culture that never seemed to fit my ideas about the need for conflict resolution was that of the Samoans. According to Margaret Mead in *Coming of Age in Samoa*, these people put an end to their quarrels by merely separating: "Disagreements between parents and child are settled by the child's moving across the street, between a man and his village by the man's removal to the next village." The anthropologist also depicted the culture as exceptionally peaceful and easygoing. We know now that this was a romantic fiction. Derek Freeman's criticism of Mead's study, based on his own intensive observations of the same people, leaves no doubt that Samoans have ties that bind, and that they, like people everywhere, try to overcome conflict rather than run away from it.

While there are as many different peacemaking rituals as there are human cultures, they all serve to turn a situation that could lead to a spiral of revenge into a mutually profitable relationship. Hostilities that have not been reconciled are stored in our memory as if packed in ice. The recollection stays fresh and cold, as we wait for an opportunity to get even. The extreme is a system of blood feud, a pattern of killing back and forth that can go on for generations. Killing can also be part of peace negotiations. The Kiwai-Papuans, for instance, may reject a peace offer of the sort described above by placing a bundle of small tally sticks on the village path to indicate how many enemy lives they intend to take before talking peace. Retaliatory reciprocity—an eye for an eye, a tooth for a tooth—is as common in humans as the cooperative reciprocity that takes its place after a peace has been sealed. When trade, intercommunity marriages, and joint feasts are resumed, we say that the incidents are "forgotten." This is, of course, nonsense. The conflict is just filed in a different, somewhat warmer pigeonhole.

Cooperative forms of reciprocity have received far more attention from science than their obverse. In connection with conflict resolution retaliation is crucial, however, as it provides the basis for our sense of justice. My data on chimpanzees demonstrate that they too keep negative acts in mind, repaying them with other negative acts. Such a "revenge system" has as yet not been found in any other animal. Humans go a step farther by setting standards—called *laws*—designed to keep feuding under control. Among the Masai, pastoral nomads of East Africa, a murderer is usually concealed by his relatives while the victim's family searches for him to avenge their kinsman's death. The culprit is protected until the mood has cooled and negotiations can begin. The traditional fine for murder is forty-nine cattle. When people collect their blood property, they go armed as though for war. Thus, the deeply rooted need for revenge is satisfied by a symbolic show of anger, punishment of the aggressor, and compensation to the victim's relations. Since it is society which sets the rules, this is a higher form of conflict settlement, not found among animals. Courts and justices are further refinements of this principle.

The legal profession may or may not represent an improvement. Certainly its original functioning was a necessity, but with an incredible 675,000 practicing lawyers in the United States (the estimate for 1985), conflicts are inevitably *created* for the sake of business. One of the culture shocks a European immigrant experiences is the inclination of Americans to put conflict resolution into the hands of attorneys and authorities. Who would call the police to report the following theft, authorize a search warrant, and name a suspect? Stolen: a rattle, a little orange car, and a stuffed gray mouse. Victim: a three-year-old boy. Suspect: his playmate of the same age. Calling on law enforcement to prosecute such a "crime," as reported in my local newspaper, may evoke a chuckle—for many Americans also—yet at the same time it indicates a sad incapacity to settle disputes.

It remains to be established whether conflict-resolution skills are weakly developed among Americans, compared to other cultural groups; but inasmuch as both the number of lawyers per capita and the murder rate are several times higher than in other industrialized nations, it seems a fair assumption. Of course, not all violent crimes can be explained this way, but a large proportion of homicides result from arguments among family members, lovers, friends, acquaintances, or neighbors. I do feel, therefore, that a nation's murder rate is inversely related to the ability of its citizens to find mutually agreeable solutions to social conflict. Few Americans deny that in their language the word "reconciliation" is almost synonymous with "capitulation." The search for compromise is not regarded as a fine art; it has a connotation of weakness.

Is toughness valued so highly in this country because historically every man had to fend for himself? Is there a connection with the belief of the early immigrants in a punishing rather than a forgiving God? Or was it the endless empty horizons and traditional mobility? At least in the past, as one American friend put it, "if people had problems, they could always move west." With so much land available, the skill of coexistence among people with different views and backgrounds may have been neglected for several generations.

The opposite is the case in my homeland. The Netherlands has one of the world's highest population densities, and "tolerance" is the key to the national character. Not that the Dutch themselves pretend that they are as tolerant as they might be, but outsiders notice a remarkable degree of noninterference with other people's business, a ready acceptance of minority life-styles and religions, and a desire for consensus.

Although the Dutch are as competitive a people as any other, their "captive" situation has promoted a different mind-set during conflict, one bent on accommodation. The Netherlands is a tiny country and, except for the cold North Sea, people have nowhere to go. That their tolerance is not a genetic trait is evident from the behavior of the Afrikaners who rule South Africa. These people are of Dutch descent, but different circumstances have led to different attitudes.

The Dutch provide a good example of a culture that contradicts the crowding theories of the sixties. Mechanisms for tension regulation upset the supposed link between crowding and aggression, as we have seen that they do in monkeys and apes. Human societies bend the rules even more, sometimes achieving completely opposite results. An abundance of space may create rugged individualists with little patience for people who stand in their way, whereas space limitations and ethnic homogeneity may lead to a collectivistic culture like that of the Japanese, with their paper-walled houses, rules of politeness, and emotional control. And all this within a single species!

The Oath of the Elbe

On November 26, 1983, the body of Joseph Polowsky, a Chicago taxi driver, was carried to a grave in Torgau, East Germany, near the Elbe River. It was there that Polowsky, as an American infantryman, had been part of the linkup between the American and Russian armies that had ended the resistance of Hitler's troops. In the intervening thirty-eight years relations between the two former allies had turned cold and hostile, but Polowsky had crafted an Oath of the Elbe, pledging to keep the wartime spirit of fraternization alive. This one-man campaign to build friendship ended with Polowsky's death from cancer and culminated in his burial far from home, in a ceremony during which both U.S. and Soviet soldiers laid wreaths in his memory.

In 1942 Nobuo Fujita, a World War II pilot of the Imperial Japanese Navy, unsuccessfully tried to ignite the forests surrounding Brookings, Oregon, by dropping incendiary bombs from his small seaplane. Two decades later the town invited Fujita to be an honorary guest at its azalea festival. By then he had become a wealthy businessman. He accepted the invitation and returned the conciliatory gesture by inviting local youngsters to Japan. Subsequently, however, Fujita lost his fortune and had to save for many years to fulfill his promise. In 1985, at the age of seventy-three, he was finally able to finance the visit of three Brookings High School students to his country. "After they have toured Japan," he told a reporter, "the war will finally be over for me."

The two stars of ethology, Niko Tinbergen and Konrad Lorenz, were on opposite sides during World War II. Tinbergen, imprisoned by the Germans during their occupation of the Netherlands, spent years in a hostage camp despite Lorenz's efforts to secure his release. Lorenz himself ended up in a Russian prison, after having served in the German army as a medical officer. In 1949, at the Cambridge home of Englishman William Thorpe, Lorenz and Tinbergen were reunited after a separation of ten years. Thorpe describes it as a moving event, but fails to provide the behavioral details one would expect from an ethologist. He does point out that just after the war Tinbergen could not stand the sound of the German language, but that this did not affect "the depth and strength of his desire for international reconciliation at all levels."

Peace efforts at the people's level may have less dramatic impact than President John F. Kennedy's "Ich bin ein Berliner" or other gestures made by leaders, yet grass-roots feelings are, in the long run, perhaps more important. The way citizens of two countries feel about one another affects cultural exchanges, business contacts, international friendships, the tone of television documentaries, and the attitude of elected officials. Although hard to predict, these mechanisms are perfectly understandable. We do not need to invoke extrasensory communication or other supernatural phenomena, as certain peace advocates do. These individuals contend that a mere *desire* for peace, if present in enough people, has the power to change the world. This belief underlies Ken Keyes' best-seller, *The Hundredth Monkey*, a volume that I need to address, as it is based on an incorrect interpretation of the primate data.

The book refers to the pioneering studies of Kinji Imanishi, Masao Kawai, and others on cultural transfer among primates. If "culture" is defined as the spreading of new habits through imitation and learning, it is perhaps quite common among animals. Potato washing by Japanese macaques was the first known example. A juvenile female named Imo discovered a way of cleaning sweet potatoes before eating them. She simply washed the sand off in the sea. Imo's peers and kin followed her example, and the behavior spread through the monkey group. Every step of the process was recorded by field-workers.

Lyall Watson read their detailed reports and in 1979 wrote a few pages concerning "the hundredth monkey phenomenon," which were then read by Keyes, who turned it into a book, which was then used by others for a film, and so on. A peculiar second part has been appended to the story. Once a certain number of monkeys had learned the habit, the addition of one more individual attained a mass of, say, one hundred monkeys. Then, we are told, it happened. From that moment the behavior suddenly spread to other populations, even to monkeys on different islands! Keyes's conclusion: "Thus, when a certain critical number achieves an awareness, this new awareness may be communicated from mind to mind." The rest of his book advocates collective consciousness of the need for a nuclear-free world. The idea is that any person tuning in to this awareness may be the one who triggers a quantum jump in public opinion.

I have no problem with use of the imagination to put across a message, but that message should not then be presented as a scientifically based truth. There is no evidence whatsoever for the second part of the monkey story (even the first part has recently become controversial). Habits do not leap across natural boundaries. The Japanese researchers all along have stressed the smoothness of the process; a breakthrough in consciousness was never observed. Ron Amundson, after careful review of the relevant data, concludes that "Watson's description of the event is refuted *in great detail* by the very sources he cites to validate it" (italics in original). He accuses Watson and the others of pseudoscience.

Confronted with the mystification of peace by some groups and the glorification of violence by others, we need to keep our heads cool. Nothing is to be gained by pacifist rallies or endless arms negotiations if there are no common interests among the powers of the world, or if there is a stubborn refusal to develop such interests. In 1987 we entered an era of optimism after U.S.

President Ronald Reagan and Soviet Premier Mikhail Gorbachev signed their historic treaty to eliminate intermediate-range missiles. There is talk of more drastic reductions to come. Yet in the absence of improved relations, nuclear disarmament could have a very disturbing effect. Some military experts foresee a buildup in conventional weapons and troops in Western Europe to offset the Soviet superiority in this area. With public opinion focusing almost exclusively on the terrible dangers of nuclear weapons, we should not forget that the value of arms treaties is limited. As the two superpowers send their armies to all corners of the world to prevent each other from gaining control, it is evident that international tensions are caused in the first place by mutual distrust and clashing ambitions. Arms are only symptoms of the disease.

The images of the Washington summit—of personal accord and joking between the two world leaders, of Gorbachev's handshaking with pedestrians, of an unprecedented visit to the Pentagon by a Soviet chief of staff—convey the impression of a fundamental change in the chemistry between the two superpowers. It is crucial that this new détente translate into expanded economic and cultural exchanges. If my studies of monkeys and apes contain any lesson for the global arena, it is that parties who need each other for one reason or another are less likely to fight; and if they do fight, they are more likely to make up afterward. If, on the other hand, a sound basis for the relationship is lacking, I am convinced that the two sides will fight regardless of the size and condition of their teeth. West German President Richard von Weizsäcker once voiced a similar view of East-West relations: "Experience teaches that it is not disarmament that points the way to peace, but rather that peaceful relations open the door to disarmament. Peace is the consequence of practical cooperation."

No doubt differences are most predictably overcome in the face of a common enemy. There are countless primate examples of this mechanism, some even involving self-created enemies. After a massive conflict in the Arnhem chimpanzee colony, while the participants were still catching their breath, one of them started aggressive *wraaa* calls in the direction of the adjacent cheetah enclosure. Others joined in; the result was a noisy, very indignant-sounding chorus of threats to the neighbors. The apes normally never paid attention to the cheetahs, and this time the cats were not even visible, as they had traveled to the far corner of their large park. In the wake of this discharge of tension a number of reconciliations took place among the chimpanzees.

In a similar situation I have seen long-tailed macaques run to their swimming pool to threaten their own images in the water; a dozen tense monkeys unified against the "other" group in the pool. The need for a common enemy can be so great that a substitute is fabricated. If such invention is not necessary because of the presence of a suitable target, internal tensions may stir up external relationships. According to Hans Kummer, battles between different bands of wild hamadryas baboons often start when members of one band "solve" a dispute among themselves by jointly threatening the members of another band.

When Christian Welker tried to establish a captive breeding colony of capuchin monkeys, he encountered serious problems, as had others before him. By trial and error, with successive introductions of monkeys, he discovered that capuchins organize themselves into subgroups and that peaceful coexistence requires these subgroups to be in balance. If one subgroup has nothing to fear from the other, violent fighting erupts—not only between the subgroups but especially within the stronger subgroup. Apparently, old rivalries among subgroup members flare up as soon as these individuals have the upper hand in the colony. When the balance is restored, either by adding members to the weaker subgroup or by removing members from the stronger, friendly contacts are resumed both within and between the subgroups. Welker speaks of the capuchin monkey's "ability to suppress enmities, and its inability to forget them."

Is this not exactly what happens at the international level? The Western allies behave as friends most of the time, yet when the American president in 1985 visited the German war cemetery at Bitburg in order to heal old wounds, some felt he had only opened them. Also, the outcry concerning the election of Austrian President Kurt Waldheim, accused of a Nazi past, shows that history is not forgotten. In today's world many former enemies have overcome their differences for reasons of national security, resulting in two blocs of "friends" each of which will stay together only as long as the other bloc remains strong.

The big puzzle of course is how these blocs can be reconciled in the absence of a common enemy. Perhaps the threat of nuclear war is taking the place of the alien invaders who might otherwise do the job. If the prospect of an unwinnable war will not bring the human species to its senses, nothing will. The capacity to foresee the consequences of our actions has helped us to plan countless wars; it may help us now to plan a future without war. The process would certainly be stimulated by the development of joint ventures, such as the proposal to organize a U.S.-Soviet trip to Mars, a gigantic enterprise. As Carl Sagan has commented, this would be a fitting symbol on behalf of humanity: "We should embrace not the god of war, but the planet named after him."

For a safe, peaceful world much, much more than arms reduction is needed. We must teach our children new goals, different skills, and global responsibility. They need to learn that their country's flag is no more than a symbol of the particular cultural group to which they belong; it does not stand for superiority, nor should they rally behind it for other offensive reasons. They should also learn that winning is only one form of conflict resolution; compromise is another, no less honorable way. I believe that such things can be taught, that the human species has in place all the mechanisms for a reorientation from confrontation to negotiation.

Conclusion

The message of this book is at odds with that of some biologists, who have one-sidedly emphasized the aggressive nature of our species and the ruthless struggle within the animal kingdom.

Ever since Darwin, the biological spotlight has been on the outcome of competition—who wins, who loses. When social animals are involved, this is a dreadful simplification. Antagonists do more than estimate their chances of winning before they engage in a fight: they also take into account how much they need their opponent. The contested resource often is simply not worth putting a valuable relationship at risk. And if aggression does occur, both parties may hurry to repair the damage. Victory is rarely absolute among interdependent competitors, whether animal or human.

Jean-Jacques Rousseau believed that there is no evil in the human heart, that all the ills of humanity began with civilization. Yet aggression is one of a large number of human behavioral characteristics that cross the boundaries of language, culture, race, even species: it cannot be fully understood without taking the biological component into account. This book has, I trust, demonstrated that appropriate countermeasures evolved along with aggressive behavior, and that both humans and other primates apply these measures with great skill. The basic pattern is that two conflicting individuals or parties become friends again. The process sounds simple enough, yet it is one of the most complex transitions we can go through.

Forgiveness is not, as some people seem to believe, a mysterious and sublime idea that we owe to a few millennia of Judeo-Christianity. It did not originate in the minds of people and cannot therefore be appropriated by an ideology or a religion. The fact that monkeys, apes, and humans all engage in reconciliation behavior means that it is probably over thirty million years old, preceding the evolutionary divergence of these primates. The alternative explanation, that this behavior appeared independently in each species, is highly "uneconomical," for it requires as many theories as there are species. Scientists normally dismiss uneconomical explanations unless there is strong evidence against the more elegant unified theory. Because no such evidence exists in this instance, reconciliation behavior must be seen as a shared heritage of the primate order. Our species has many conciliatory gestures and contact patterns in common with the apes (stretching out a hand, smiling, kissing, embracing, and so on). Language and culture merely add a degree of subtlety and variation to human peacemaking strategies.

This knowledge does not solve the problem of violence in our societies, but I do hope that it will bring a change in perspective. Instead of looking at reconciliation as a triumph of reason over instinct, we need to begin to study the roots and universality of the psychological mechanisms involved. It is time for science to enter the scene. A rational approach should replace the mystique that surrounds the peace issue today. We need not be under the illusion that aggressive tendencies will ever leave us, but neither should we neglect our heritage of reconciliation. In shifting the emphasis from the one to the other, we would in no way be crossing the boundaries of human nature. We would only be making use of what we have, and doing what we do best—adapting to new circumstances in our own self-interest.

Bibliography

Altmann, S. 1981. Dominance relationships: the Cheshire cat's grin? *Behav. Brain Sci.* 4:430–431.

Amundson, R. 1985. The hundredth monkey phenomenon. *Sceptical Enquirer* 9:348–356.

Aronson, E. 1976. *The Social Animal,* 2nd ed. San Francisco: Freeman.

Artaud, Y., and M. Bertrand. 1984. Unusual manipulatory activity and tool-use in a crab-eating macaque. In M. Roonwal et al., eds., *Current Primate Researches.* Jodhpur: University of Jodhpur Press.

van den Audenaerde, D. 1984. The Tervuren Museum and the pygmy chimpanzee. In R. Susman, ed., *The Pygmy Chimpanzee,* pp. 3–11. New York: Plenum.

Bachmann, C., and H. Kummer. 1980. Male assessment of female choice in hamadryas baboons. *Behav. Ecol. Sociobiol.* 6:315–321.

Badrian, A. 1984. The bonobo branch of the family tree. *Anim. Kingdom* 87:39–45.

Badrian, A., and N. Badrian. 1984. Social organization of *Pan paniscus* in the Lomako Forest, Zaire. In R. Susman, ed., *The Pygmy Chimpanzee,* pp. 325–346. New York: Plenum.

Bahn, C. 1980. Hostage taking—the takers, the taken, and the context: discussion. *Ann. NY Acad. Sci.* 347:151–156.

Bandura, A., D. Ross, and S. Ross. 1961. Transmission of aggression through imitation of aggressive models. *J. Abn. Soc. Psychol.* 63:575–582.

Barash, D. 1977. *Sociobiology and Behavior.* New York: Elsevier.

Bauman, J. 1926. Observations on the strength of the chimpanzee and its implications. *J. Mammal.* 7:1–9.

Beck, B. 1982. Chimpocentrism: bias in cognitive ethology. *J. Human Evol.* 11:3–17.

Becker, C. 1983. Sozialspiel in einer gemischten Gruppe Orang-utans und Bonobos, sowie Spielverhalten aller Orang-utans im Kölner Zoo. *Z. Kölner Zoo* 26:59–69.

Bernstein, I., L. Williams, and M. Ramsay. 1983. The expression of aggression in Old World monkeys. *Int. J. Primatol.* 4:113–125.

Bertrand, M. 1969. *The Behavioral Repertoire of the Stumptail Macaque.* Bibliotheca Primatologica, vol. 11. Basel: Karger.

Bleier, R. 1984. *Science and Gender.* New York: Pergamon.

Bohannan, P. 1983. Some bases of aggression and their relationship to law. In M. Gruter and P. Bohannan, eds., *Law, Biology and Culture,* pp. 147–158. Santa Barbara, Calif.: Ross-Erikson.

Bolk, L. 1926. *Das Problem der Menschwerdung.* Jena: Gustav Fischer Verlag.

Bond, J., and W. Vinacke. 1961. Coalitions in mixed-sex triads. *Sociometry* 24:61–75.

van Bree, P. 1963. On a specimen of *Pan paniscus,* Schwarz 1929, which lived in the Amsterdam Zoo from 1911 till 1916. *Zool. Garten* 27:292–295.

Brehm, A. 1916. *Brehms Tierleben: Algemeine Kunde des Tierreichs,* vol. 13, 4th ed. Leipzig: Bibliographisches Institut.

Bygott, J. D. 1974. Agonistic behavior and dominance in wild chimpanzees. Ph.D. diss., Cambridge University.

Campbell, S. 1980. Kakowet. *Zoonooz* 53:6–11.

Caplow, T. 1968. *Two against One: Coalitions in Triads.* Englewood Cliffs, N.J.: Prentice-Hall.

Chagnon, N. 1968. *Yanomamö: The Fierce People.* New York: Holt, Rinehart and Winston.

Chance, M. 1967. Attention structure as the basis of primate rank orders. *Man* 2:503–518.

Cheney, D., and R. Seyfarth. 1986. The recognition of social alliances by vervet monkeys. *Anim. Behav.* 34:1722–31.

Clavell, J. 1981. *Noble House.* Philadelphia, Penn.: Coronet Books.

Coe, C., and L. Rosenblum. 1984. Male dominance in the bonnet macaque: a malleable relationship. In P. Barchas and S. Mendoza, eds., *Social Cohesion,* pp. 31–63. Westport, Conn.: Greenwood Press.

Coolidge, H. 1933. *Pan paniscus:* pygmy chimpanzee from south of the Congo River. *Am. J. Phys. Anthrop.* 18:1–57.

——— 1984. Historical remarks bearing on the discovery of *Pan paniscus.* In R. Susman, ed., *The Pygmy Chimpanzee,* pp. ix–xiii. New York: Plenum.

Cruise O'Brien, C. 1984. Religions, cultures and conflict. In P. Dorner, ed., *World without War.* Madison: Office of International Studies and Programs, University of Wisconsin.

Cummings, E. M., C. Zahn-Waxler, and M. Radke-Yarrow. 1981. Young children's responses to expressions of anger and affection by others in the family. *Child Development* 52:1274–82.

Curie-Cohen, M., et al. 1983. The effects of dominance on mating behavior and paternity in a captive group of rhesus monkeys. *Am. J. Primatol.* 5:127–138.

Dahl, J. 1985. The external genitalia of female pygmy chimpanzees. *Anat. Rec.* 211:24–48.

——— 1986. Cyclic perineal swelling during the intermenstrual intervals of captive female pygmy chimpanzees. *J. Human Evol.* 15:369–385.

Darwin, C. 1859. *The Origin of Species.* London: John Murray.

Dasser, V. 1988. A social concept in Java-monkeys. *Anim. Behav.* 36:225–230.

Dawkins, R. 1976. *The Selfish Gene.* New York: Oxford University Press.

Diamond, J. 1984. DNA map of the human lineage. *Nature* 310:544.

Dittus, W. 1979. The evolution of behaviors regulating density and age-specific sex ratios in a primate population. *Behaviour* 69:265–302.

Eckholm, E. 1985. Pygmy chimp readily learns language skills. *New York Times*, June 25.

Eibl-Eibesfeldt, I. 1971 (1970). *Love and Hate.* New York: Holt, Rinehart and Winston.

——— 1976. *Der vorprogrammierte Mensch.* Munich: Deutscher Taschenbuch Verlag.

——— 1977. Patterns of greeting in New Guinea. In S. Wurm, ed., *Language, Culture, Society, and the Modern World.* Canberra: Australian National University Press.

——— 1980. Strategies of social interaction. In R. Plutchnik and H. Kellerman, eds., *Theories of Emotion.* New York: Academic Press.

Ekman, P. 1982. *Emotion in the Human Face*, 2nd ed. Cambridge: Cambridge University Press.

Elton, R. 1979. Baboon behavior under crowded conditions. In J. Erwin, T. Maple, and G. Mitchell, eds., *Captivity and Behavior*, pp. 125–138. New York: Van Nostrand.

Erwin, J. 1979. Aggression in captive macaques: interaction of social and spatial factors. In J. Erwin, T. Maple, and G. Mitchell, eds., *Captivity and Behavior*, pp. 139–171. New York: Van Nostrand.

Ettlinger, G. 1984. Comment. In R. Harré and V. Reynolds, eds., *The Meaning of Primate Signals*, pp. 109–110. Cambridge: Cambridge University Press.

Fallaci, O. 1976. *Interview with History.* Boston: Houghton Mifflin.

Fedigan, L. 1983. Dominance and reproductive success in primates. *Yearb. Phys. Anthrop.* 26:91–129.

Fisher, H. 1983. *The Sex Contract: The Evolution of Human Behavior.* New York: Quill.

Fooden, J., et al. 1985. The stumptail macaques of China. *Am. J. Primatol.* 8:11–30.

Ford, C., and F. Beach. 1951. *Patterns of Sexual Behavior.* New York: ACE Books.

Fossey, D. 1983. *Gorillas in the Mist.* Boston: Houghton Mifflin.

Fox, M. 1982. Are most animals "mindless automatons"? A reply to Gordon G. Gallup, Jr. *Am. J. Primatol.* 3:341–343.

Freeman, D. 1983. *Margaret Mead and Samoa.* Cambridge, Mass.: Harvard University Press.

French, M. 1985. *Beyond Power.* New York: Ballantine.

von Frisch, K. 1923. Über die "Sprache" der Bienen. *Zool. Jahrb. Abt. allg. Zool. Physiol. Tiere* 40:1–186.

Gallup, G. 1982. Self-awareness and the emergence of mind in primates. *Am. J. Primatol.* 2:237–248.

Gerard, H., and G. Mathewson. 1966. The effects of severity of initiation on liking for a group: a replication. *J. Exp. Soc. Psychol.* 2:278–287.

Ginsburg, H. 1980. Playground as laboratory: naturalistic studies of appeasement, altruism and the Omega child. In D. Omark, F. Strayer, and D. Freedman, eds., *Dominance Relations,* pp. 341–357. New York: Garland.

Goldfoot, D., et al. 1980. Behavioral and physiological evidence of sexual climax in the female stump-tailed macaque. *Science* 208:1477–79.

Golding, W. 1954. *Lord of the Flies.* New York: Capricorn.

Goldstein, J. 1986. *Aggression and Crimes of Violence,* 2nd ed. New York: Oxford University Press.

Goodall, J. 1971. *In the Shadow of Man.* London: Collins.

——— 1983. Population dynamics during a 15-year period in one community of free-living chimpanzees in the Gombe National Park, Tanzania. *Z. Tierpsychol.* 61:1–60.

——— 1986a. Social rejection, exclusion and shunning among the Gombe chimpanzees. *Ethol. Sociobiol.* 7:227–236.

——— 1986b. *The Chimpanzees of Gombe.* Cambridge, Mass.: Belknap Press, Harvard University Press.

Goodall, J., et al. 1979. Intercommunity interactions in the chimpanzee population of the Gombe National Park. In D. Hamburg and E. McCown, eds., *The Great Apes*, pp. 13–53. Menlo Park, Calif.: Benjamin/Cummings.

Gould, S. 1977. *Ontogeny and Phylogeny.* Cambridge, Mass.: Belknap Press, Harvard University Press.

Gribbin, J., and J. Cherfas. 1982. *The Monkey Puzzle.* New York: Pantheon.

Griede, T. 1981. Invloed op verzoening bij chimpansees. Research report, University of Utrecht.

Guinness, A. 1986. *Blessings in Disguise.* New York: Knopf.

Hahn, E. 1982. Annals of zoology; a moody giant. *New Yorker,* August.

Halperin, S. 1979. Temporary association patterns in free-ranging chimpanzees. In D. Hamburg and E. McCown, eds., *The Great Apes,* pp. 491–499. Menlo Park, Calif.: Benjamin/Cummings.

Hand, J. 1986. Resolution of social conflicts: dominance, egalitarianism, spheres of dominance, and game theory. *Q. Rev. Biol.* 61:201–220.

Hardy, A. 1960. Was man more aquatic in the past? *New Scientist* 7:642–645.

Harlow, H., and M. Harlow. 1965. The affectional systems. In A. Schrier, H. Harlow, and F. Stollnitz, eds., *Behavior of Nonhuman Primates,* vol. 2, pp. 287–334. New York: Academic Press.

Harlow, H., and C. Mears. 1979. *The Human Model.* New York: Wiley.

Herschberger, R. 1948. *Adam's Rib.* New York: Harper and Row.

Heublein, E. 1977. Kakowet's family. *Zoonooz* 50:4–10.

Heuvelmans, B. 1980. *Les bêtes humaines d'Afrique.* Paris: Plon.

Hibbert, C. 1965. *The Rise and Fall of Il Duce.* London: Penguin Books.

van Hooff, J. 1972. A comparative approach to the phylogeny of laughter and smiling. In R. Hinde, ed., *Non-verbal Communication,* pp. 209–241. Cambridge: Cambridge University Press.

Horn, A. 1976. A preliminary report on the ecology and behavior of the bonobo chimpanzee, and a reconsideration of the evolution of the chimpanzee. Ph.D. diss., Yale University.

Hrdy, S. 1981. *The Woman That Never Evolved.* Cambridge, Mass.: Harvard University Press.

Huizinga, J. 1972 (1950). *Homo ludens: A Study of the Play-Element in Culture.* Boston: Beacon Press.

Huxley, T. H. 1888. Struggle for existence and its bearing upon man. *Nineteenth Century,* February.

Imanishi, K. 1965 (1957). Identification: a process of socialization in the subhuman society of *Macaca fuscata. Primates* 1:1–29. English translation in K. Imanishi and S. Altmann, eds., *Japanese Monkeys*. Atlanta: Emory University.

Itani, J., and A. Mishimura. 1973. The study of infrahuman culture in Japan: a review. In E. Menzel, ed., *Precultural Primate Behavior*, pp. 26–50. Basel: Karger.

Jordan, C. 1977. Das Verhalten zoolebender Zwergschimpansen. Ph.D. diss., Goethe University, Frankfurt.

Jungers, W., and R. Susman. 1984. Body size and skeletal allometry in African apes. In R. Susman, ed., *The Pygmy Chimpanzee*, pp. 131–177. New York: Plenum.

Kano, T. 1979. A pilot study on the ecology of pygmy chimpanzees. In D. Hamburg and E. McCown, eds., *The Great Apes*, pp. 123–135. Menlo Park, Calif.: Benjamin/Cummings.

———— 1984a. Distribution of pygmy chimpanzees in the Central Zaire Basin. *Folia Primatol.* 43:36–52.

———— 1984b. Observations of physical abnormalities among the wild bonobos of Wamba, Zaire. *Am. J. Phys. Anthrop.* 63:1–11.

Kano, T., and M. Mulavwa. 1984. Feeding ecology of the pygmy chimpanzees of Wamba. In R. Susman, ed., *The Pygmy Chimpanzee*, pp. 233–274. New York: Plenum.

Kaplan, J. 1978. Fight interference and altruism in rhesus monkeys. *Am. J. Phys. Anthrop.* 49:241–250.

Kaufman, I. 1974. Mother/infant relations in monkeys and humans: a reply to Professor Hinde. In N. White, ed., *Ethology and Psychiatry*, pp. 47–68. Toronto: University of Toronto Press.

Kawai, M. 1965 (1958). On the system of social ranks in a natural troop of Japanese monkeys. *Primates* 1:111–148. English translation in K. Imanishi and S. Altmann, eds., *Japanese Monkeys*. Atlanta: Emory University.

———— 1965. On the newly acquired pre-cultural behavior of the natural troop of Japanese monkeys on Koshima Islet. *Primates* 6:1–30.

Kawamura, S. 1965 (1958). Matriarchal social ranks in the Minoo-B troop: a study of the rank system of Japanese monkeys. *Primates* 1:148–156. English translation in K. Imanishi and S. Altmann, eds., *Japanese Monkeys*. Atlanta: Emory University.

Kawanaka, K. 1984. Association, ranging, and the social unit in chimpanzees of the Mahale Mountains, Tanzania. *Int. J. Primatol.* 5:411–432.

Kellogg, W., and L. Kellogg. 1933. *The Ape and the Child*. New York: McGraw-Hill.

Keyes, K. 1982. *The Hundredth Monkey*. Coos Bay, Oreg.: Vision Books.

King, M., and A. Wilson. 1975. Evolution at two levels in humans and chimpanzees. *Science* 188:107–116.

Kling, A., and J. Orbach. 1963. The stump-tailed macaque: a promising laboratory primate. *Science* 139:45–46.

Köhler, W. 1925. *The Mentality of Apes*. New York: Vintage Books.

Kornfeld, A. 1975. *In a Bluebird's Eye*. New York: Avon Books.

Kortlandt, A. 1976. Statements on pygmy chimpanzees. *Lab. Primate Newsletter* 15:15–17.

Kummer, H. 1957. *Soziales Verhalten einer Mantelpavian-Gruppe*. Bern: Huber.

———— 1968. *Social Organization of Hamadryas Baboons*. Chicago: University of Chicago Press.

———— 1984. From laboratory to desert and back: a social system of hamadryas baboons. *Anim. Behav.* 32:965–971.

Kummer, H., W. Götz, and W. Angst. 1974. Triadic differentiation: an inhibitory process protecting pair bonds in baboons. *Behaviour* 49:62–87.

Kuroda, S. 1984. Interaction over food among pygmy chimpanzees. In R. Susman, ed., *The Pygmy Chimpanzee*, pp. 301–324. New York: Plenum.

Landtman, G. 1927. *The Kiwai Papuans of British New Guinea*. London: Macmillan.

Lefebvre, L. 1982. Food exchange strategies in an infant chimpanzee. *J. Human Evol.* 11:195–204.

Lethmate, J., and Ducker, G. 1973. Untersuchungen zum Selbsterkennen im Spiegel bei Orang-Utans und einigen anderen Affenarten. *Z. Tierpsychol.* 33:248–269.

Lindburg, D. 1971. The rhesus monkey in North India: an ecological and behavioral study. In L. Rosenblum, ed., *Primate Behavior*, pp. 2–106. New York: Academic Press.

Linnankoski, I., and L. Leinonen. 1985. Compatibility of male and female sexual behaviour in *Macaca arctoides*. *Z. Tierpsychol.* 70:115–122.

Lorenz, K. 1967 (1963). *On Aggression*. London: Methuen.

———— 1981. *The Foundations of Ethology*. New York: Simon and Schuster.

Lovejoy, C. O. 1981. The origin of man. *Science* 211:341–350.

Maccoby, E., and C. Jacklin. 1974. *The Psychology of Sex Differences.* Stanford: Stanford University Press.

Machiavelli, N. 1979 (1532). *The Prince.* In P. Bondanella and M. Musa, eds., *The Portable Machiavelli,* pp. 47–166. Harmondsworth: Penguin Books.

MacKinnon, J. 1978. *The Ape within Us.* London: Collins.

Malinowski, B. 1922. *Argonauts of the Western Pacific.* London: Routledge and Kegan Paul.

Mason, W. 1965. Determinants of social behavior in young chimpanzees. In A. Schrier, H. Harlow, and F. Stollnitz, eds., *Behavior of Nonhuman Primates,* vol. 2, pp. 335–364. New York: Academic Press.

Masserman, J., S. Wechkin, and W. Terris. 1964. "Altruistic" behavior in rhesus monkeys. *Am. J. Psychiatry* 121:584–585.

Massey, A. 1977. Agonistic aids and kinship in a group of pigtail macaques. *Behav. Ecol. Sociobiol.* 2:31–40.

Masters, W., and V. Johnson. 1966. *Human Sexual Response.* Boston: Little, Brown.

Mayer, C. 1960. *Caste and Kinship in Central India.* London: Routledge and Kegan Paul.

McGuire, M., M. Raleigh, and C. Johnson. 1983. Social dominance in adult male vervet monkeys: general considerations. *Soc. Sci. Information* 22:89–123.

Mead, M. 1943 (1928). *Coming of Age in Samoa.* Harmondsworth: Penguin Books.

Melnick, D., and K. Kidd. 1985. Genetic and evolutionary relationships among Asian macaques. *Int. J. Primatol.* 6:123–160.

Milgram, S. 1974. *Obedience to Authority.* New York: Harper and Row.

Montagner, H. 1978. *L'enfant et la communication.* Paris: Stock.

Montagu, A., ed. 1968. *Man and Aggression.* London: Oxford University Press.

Morgan, E. 1982. *The Aquatic Ape.* New York: Stein and Day.

Mori, A. 1984. An ethological study of pygmy chimpanzees in Wamba, Zaire: a comparison with chimpanzees. *Primates* 25:255–278.

Morris, D. 1967. *The Naked Ape.* London: Jonathan Cape.

Morrow, L. 1984. I spoke as a brother: a pardon from the pontiff, a lesson in forgiveness for a troubled world. *Time,* January 9.

Mussolini, B. 1939. *My Autobiography.* London: Hutchinson.

Myers, G. 1972 (1949). A monograph on the piranha. In G. Myers, ed., *The Piranha Book.* Neptune City: Tropical Fish Hobbyist Publications.

Nacci, P., and J. Tedeschi. 1976. Liking and power as factors affecting coalition choices in the triad. *Soc. Behav. Person.* 4:27–31.

Napier, J. 1975. The talented primate. In V. Goodall, ed., *The Quest for Man.* London: Phaidon.

Nieuwenhuijsen, K. 1985. Geslachtshormonen en gedrag bij de beermakaak. Ph.D. diss., Erasmus University, Rotterdam.

Nieuwenhuijsen, K., and F. de Waal. 1982. Effects of spatial crowding on social behavior in a chimpanzee colony. *Zoo Biology* 1:5–28.

Nishida, T. 1979. The social structure of chimpanzees in the Mahale Mountains. In D. Hamburg and E. McCown, eds., *The Great Apes,* pp. 73–121. Menlo Park, Calif.: Benjamin/Cummings.

———— 1983. Alpha status and agonistic alliance in wild chimpanzees. *Primates* 24:318–336.

———— Forthcoming. Social structure and dynamics of chimpanzees: a review. In P. Seth and S. Seth, eds., *Perspectives in Primate Biology.*

Nishida, T., et al. 1985. Group extinction and female transfer in wild chimpanzees in the Mahale National Park, Tanzania. *Z. Tierpsychol.* 67:284–301.

Nissen, H., and M. Crawford. 1936. A preliminary study of food-sharing behavior in young chimpanzees. *J. Comp. Psychol.* 22:383–419.

Nixon, R. 1983. *Real Peace: A Strategy for the West.* Privately published; quoted in *Time,* September 19, 1983.

Noë, R. 1986. Lasting alliances among adult male savannah baboons. In J. Else and P. Lee, eds., *Primate Ontogeny,* pp. 381–392. Cambridge: Cambridge University Press.

van Noordwijk, M., and C. van Schaik. 1985. Male migration and rank acquisition in wild long-tailed macaques. *Anim. Behav.* 33:849–861.

———— 1987. Competition among female long-tailed macaques, *Macaca fascicularis. Anim. Behav.* 35:577–589.

Offit, A. 1981. *Night Thoughts: Reflections of a Sex Therapist.* New York: Congdon and Lattes.

Packer, C. 1977. Reciprocal altruism in *Papio anubis. Nature* 265: 441–443.

———— 1979. Male dominance and reproductive activity in *Papio anubis. Anim. Behav.* 27:37–45.

Patterson, T. 1973. The behavior of a group of captive pygmy chimpanzees. Master's thesis, University of Georgia.

Portielje, A. 1916. *Een Gids bij den Rondgang.* Amsterdam: Natura Artis Magistra.

Premack, D., and A. Premack. 1983. *The Mind of an Ape.* New York: Norton.

Pugh, G. 1977. *The Biological Origin of Human Values.* New York: Basic Books.

Pusey, A. 1979. Intercommunity transfer of chimpanzees in Gombe National Park. In D. Hamburg and E. McCown, eds., *The Great Apes,* pp. 465–479. Menlo Park, Calif.: Benjamin/ Cummings.

Reynolds, V. 1967. On the identity of the ape described by Tulp, 1641. *Folia Primatol.* 5:80–87.

Rijksen, H. 1977. Sumatran orang utans. Ph.D. diss., Landbouw-hogeschool, Wageningen.

Riss, D., and J. Goodall. 1977. The recent rise to the alpha-rank in a population of free-living chimpanzees. *Folia Primatol.* 27:134–151.

Rubin, L. 1985. *Just Friends.* New York: Harper and Row.

Sackin, S., and E. Thelen. 1984. An ethological study of peaceful associative outcomes to conflict in preschool children. *Child Development* 55:1098–1102.

Sagan, C. 1977. *Dragons of Eden.* New York: Random House.

Sankan, S. 1971. *The Maasai.* Nairobi: Kenya Literature Bureau.

Savage-Rumbaugh, S. 1984. *Pan paniscus* and *Pan troglodytes:* contrasts in preverbal communicative competence. In R. Susman, ed., *The Pygmy Chimpanzee,* pp. 395–413. New York: Plenum.

Savage-Rumbaugh, S., and B. Wilkerson. 1978. Socio-sexual behavior in *Pan paniscus* and *Pan troglodytes:* a comparative study. *J. Human Evol.* 7:327–344.

Schenkel, R. 1967. Submission: its features and function in the wolf and dog. *Am. Zool.* 7:319–323.

Schropp, R. 1985. Children's use of objects—competitive or interactive? Paper presented at the 19th International Ethological Conference, Toulouse.

Schubert, G. 1986. Primate politics. *Soc. Sci. Information* 25:647–680.

Schwarz, E. 1929. Das Vorkommen des Schimpansen auf dem linken Kongo-Ufer. *Rev. Zool. Bot. Afr.* 16:425–426.

Scott, J. 1972 (1958). *Animal Behavior,* 2nd ed. Chicago: University of Chicago Press.

Seville statement on violence. 1986. Middletown, Conn.: Wesleyan University.

Seyfarth, R. 1977. A model of social grooming among adult female monkeys. *J. Theor. Biol.* 65:671–698.

Sibley, C., and Ahlquist, J. 1984. The phylogeny of the Hominoid primates, as indicated by DNA-DNA hybridization. *J. Mol. Evol.* 20:2–15.

Simmel, G. 1970 (1917). *Grundfragen der Soziologie.* Berlin: Walter de Gruyter.

Skinner, B. 1971. *Beyond Freedom and Dignity.* New York: Knopf.

Slob, K., et al. 1978. Heterosexual interactions in laboratory-housed stumptail macaques (*Macaca arctoides*): observations during the menstrual cycle and after ovariectomy. *Horm. Behav.* 10:193–211.

Smith, D. 1981. The association between rank and reproductive success of male rhesus monkeys. *Am. J. Primatol.* 1:83–90.

Smuts, B. 1985. *Sex and Friendship in Baboons.* New York: Aldine.

Southwick, C. 1980. Rhesus monkey populations in India and Nepal: patterns of growth, decline, and natural regulation. In M. Cohen, ed., *Biosocial Mechanisms of Population Regulation,* pp. 151–170. New Haven: Yale University Press.

Southwick, C., M. Beg, and M. Siddiqi. 1965. Rhesus monkeys in North India. In I. DeVore, ed., *Primate Behavior,* pp. 111–159. New York: Holt, Rinehart and Winston.

Spykman, N. 1964. *The Social Theory of Georg Simmel.* New York: Russell and Russell.

Strayer, F., and J. Noel. 1986. The prosocial and antisocial functions of preschool aggression: an ethological study of triadic conflict among young children. In C. Zahn-Waxler, E. Cummings, and R. Iannotti, eds., *Altruism and Aggression,* pp. 107–131. Cambridge: Cambridge University Press.

Strayer, F., and M. Trudel. 1984. Developmental changes in the nature and function of social dominance among young children. *Ethol. Sociobiol.* 5:279–295.

Susman, R. 1984. The locomotor behavior of *Pan paniscus* in the Lomako Forest. In R. Susman, ed., *The Pygmy Chimpanzee,* pp. 369–391. New York: Plenum.

Susman, R., and K. Kabonga. 1984. Update on the pygmy chimp in Zaire. *IUCN/SSC Primate Specialist Group Newsletter* 4:34–36.

Susman, R., J. Stern, and W. Jungers. 1984. Arboreality and bipedality in the Hadar hominids. *Folia Primatol.* 43:113–156.

Suzuki, A. 1971. Carnivority and cannibalism observed among forest-living chimpanzees. *J. Anthrop. Soc. Nippon* 74:30–48.

Swanson, H., and R. Schuster. 1987. Cooperative social coordination and aggression in male laboratory rats: effects of housing and testosterone. *Hormones and Behav.* 21:310–330.

Symons, D. 1978. The question of function: dominance and play. In E. Smith, ed., *Social Play in Primates*, pp. 193–230. New York: Academic Press.

—— 1979. *The Evolution of Human Sexuality.* New York: Oxford University Press.

Takahata, Y., T. Hasegawa, and T. Nishida. 1984. Chimpanzee predation in the Mahale Mountains from August 1979 to May 1982. *Int. J. Primatol.* 5:213–233.

Teas, J., et al. 1982. Aggressive behavior in the free-ranging rhesus monkeys of Kathmandu, Nepal. *Aggress. Behav.* 8:63–77.

Terrace, H. 1979. *Nim: A Chimpanzee Who Learned Sign Language.* New York: Washington Square Press.

Thierry, B. 1984. Clasping behavior in *Macaca tonkeana. Behaviour* 89:1–28.

—— 1986. A comparative study of aggression and response to aggression in three species of macaque. In J. Else and P. Lee, eds., *Primate Ontogeny, Cognition and Social Behaviour*, pp. 307–313. Cambridge: Cambridge University Press.

Thompson-Handler, N., R. Malenky, and N. Badrian. 1984. Sexual behavior of *Pan paniscus* under natural conditions in the Lomako Forest, Equateur, Zaire. In R. Susman, ed., *The Pygmy Chimpanzee*, pp. 347–368. New York: Plenum.

Thorpe, W. 1979. *The Origins and Rise of Ethology.* London: Heineman.

Tratz, E., and H. Heck. 1954. Der afrikanische Anthropoide "Bonobo," eine neue Menschenaffengattung. *Saugetierkundige Mitt.* 2:97–101.

Tulp, N. 1641. *Observationum medicarum libri tres.* Amsterdam. Cited in Reynolds, 1967.

Turnbull, C. 1962. *The Forest People.* New York: Touchstone.

Vauclair, J., and K. Bard. 1983. Development of manipulations with objects in ape and human infants. *J. Human Evol.* 12:631–645.

de Waal, F. 1975. The wounded leader: a spontaneous temporary change in the structure of agonistic relations among captive Java-monkeys (*Macaca fascicularis*). *Neth. J. Zool.* 25:529–549.

—— 1982. *Chimpanzee Politics.* London: Jonathan Cape.

—— 1984a. Coping with social tension: sex differences in the effect of food provision to small rhesus monkey groups. *Anim. Behav.* 32:765–773.

—— 1984b. Sex differences in the formation of coalitions among chimpanzees. *Ethol. Sociobiol.* 5:239–255.

—— 1986. Integration of dominance and social bonding in primates. *Q. Rev. Biol.* 61:459–479.

—— 1987. Tension regulation and nonreproductive functions of sex in captive bonobos (*Pan paniscus*). *Nat. Geogr. Research* 3:318–335.

—— Forthcoming. Reconciliation among primates: a review of empirical evidence and theoretical issues. In W. Mason and S. Mendoza, eds., *Primate Social Conflict.* New York: Alan Liss.

—— In press. The myth of a simple relation between space and aggression in captive primates. *Zoo Biology.*

de Waal, F., and L. Luttrell. 1986. The similarity principle underlying social bonding among female rhesus monkeys. *Folia Primatol.* 46:215–234.

—— 1988. Mechanisms of social reciprocity in three primate species: symmetrical relationship characteristics or cognition? *Ethol. Sociobiol.* 9:101–118.

de Waal, F., and R. Ren. 1988. Comparison of the reconciliation behavior of stumptail and rhesus macaques. *Ethology* 78:129–142.

de Waal, F., and A. van Roosmalen. 1979. Reconciliation and consolation among chimpanzees. *Behav. Ecol. Sociobiol.* 5:55–66.

de Waal, F., and D. Yoshihara. 1983. Reconciliation and redirected affection in rhesus monkeys. *Behaviour* 85:224–241.

Walters, J. 1980. Interventions and the development of dominance relationships in female baboons. *Folia Primatol.* 34:61–89.

Watson, L. 1979. *Lifetide.* New York: Simon and Schuster.

Welker, C. 1981. Zum Sozialverhalten des Kapuzineraffen (*Cebus apella*) in Gefangenschaft. *Philippia* 4:331–342.

White, L. 1959. *The Evolution of Culture.* New York: McGraw-Hill.

Wilson, E. 1975. *Sociobiology: The New Synthesis.* Cambridge, Mass.: Belknap Press, Harvard University Press.

Witt, R., C. Schmidt, and J. Schmitt. 1981. Social rank and Darwinian fitness in a multimale group of barbary macaques. *Folia Primatol.* 36:201–211.

Wrangham, R. 1979. Sex differences in chimpanzee dispersion. In D. Hamburg and E. McCown, eds., *The Great Apes*, pp. 481–490. Menlo Park, Calif.: Benjamin/Cummings.

Yerkes, R. 1925a. *Almost Human.* New York: Century.
—— 1925b. Traits of young chimpanzees. In R. Yerkes and B. Learned, eds., *Chimpanzee Intelligence and Its Vocal Expressions,* pp. 11–56. Baltimore: Williams and Wilkins.
—— 1941. Conjugal contrasts among chimpanzees. *J. Abnorm. Soc. Psychol.* 36:175–199.
York, A., and T. Rowell. 1988. Reconciliation following aggression in patas monkeys, *Erythrocebus patas. Anim. Behav.* 36:502–509.
Zihlman, A. 1984. Body build and tissue composition in *Pan paniscus* and *Pan troglodytes* with comparisons to other Hominoids. In R. Susman, ed., *The Pygmy Chimpanzee,* pp. 179–200. New York: Plenum.
Zihlman, A., and J. Lowenstein. 1983. A few words with Ruby. *New Scientist,* April 14:81–83.
Zuckerman, S. 1932. *The Social Life of Monkeys and Apes.* New York: Harcourt.

Index

For ease of identification, main entries that are names of individual primates are italicized.